A VIEW OF THE SEA

A View of the Sea

A Discussion between a Chief Engineer
and an Oceanographer about the Machinery
of the Ocean Circulation

by
HENRY
STOMMEL

PRINCETON UNIVERSITY PRESS

PRINCETON, NEW JERSEY

Library of Congress Cataloging in Publication Data will be found
on the last printed page of this book

ISBN 0-691-08458-0

This book has been composed in Linotron Century Schoolbook

Clothbound editions of Princeton University Press books are printed
on acid-free paper, and binding materials are chosen for strength
and durability. Paperbacks, although satisfactory for personal
collections, are not usually suitable for library rebinding

Printed in the United States of America by Princeton
University Press, Princeton, New Jersey

To my wife and companion of thirty-seven years,
Elizabeth

Contents

Abbreviations

CF	Coriolis force
CTD	Instrument for measuring conductivity, temperature and depth in the ocean
CVE	Characteristic velocity, east
CVN	Characteristic velocity, north
ET	Ekman transport
GV	Geostrophic velocity
GV(ref)	Geostrophic velocity in the reference layer
HPG	Horizontal pressure gradient
HPG(ref)	Horizontal pressure gradient in the reference layer
PT	Potential thickness
RV	Relative vorticity
RW	Rossby wave velocity
SVE	Sverdrup velocity, east
SVN	Sverdrup velocity, north

Foreword

Your ship is about to undertake a survey of a large triangular area in the sea south of the Azores in connection with some ideas you have about the circulation of the ocean. The expedition is to be the last of five cruises that you will make in the early 1980s to establish as precisely as you can the shape of the density stratification beneath the ocean's surface. It is part of a study that is meant to reveal an essential feature of mid-ocean circulation: the beta-spiral. It will lead to a better understanding of the internal machinery of the sea.

The Las Palmas harbor pilot has assigned the ship one of the less attractive docks, reserving ones nearer the center of town for the regular crowded tourist ships. He, the port doctor, and the customs officials have all departed. The Captain has left for an afternoon with his golf clubs. The crew and scientific party are ashore exploring the local delights and scenery.

You are sitting in the mess with the chief engineer for company. He is a relief engineer, having spent most of his life on merchant ships, and this is the first time he has taken a temporary berth on a research vessel. You are both glad to have the ship to yourselves for a little while, and grateful for the relative quiet. The main engines are shut down. The noisy blowers are turned off. It is a good time for some idle chatting, and after awhile the usual topics are exhausted.

The Chief then asks you what your survey is all about, and you know that you are faced with explaining how the machinery of the ocean circulation works. The Chief, of course, knows his engines. He can take them apart and put them together again, and he knows the function of every part. So he has a pretty shrewd idea of what explaining a machine amounts to. But giving such an explanation is a daunting experience for an oceanographer. Yes, you have some ideas about the ocean's machinery, but they are largely hypothetical. Yes, you have some measurements, but not of all the variables. And the explanation you can give that sober old Chief would not qualify you for oiler in the *Ocean*'s engine room. It is futile to toss out fancy words like "geophysical fluid dynamics" to the Chief. He has met smart alecks before. There is no getting around the immense discrepancy between knowing how a machine works (as he knows the running of the ship's engines) and being at an early stage of exploring an unknown machine (as happens to be the fate of an oceanographer).

This book, then, is my attempt to explain our ideas about the physical machinery of the ocean's circulation to that mildly curious old Chief.

The reader will find some computer programs listed in an appendix at the end of the book. They can be run on most personal computers with color-graphics capability, and are meant to give some extra "hands-on" experience with the ideas on ocean circulation presented in this book. They are not trivial games but useful, optional tools for the serious reader. The text of the book, however, is written to stand on its own, and the programs are not required for an understanding of the contents of this book. Those who wish to purchase the programs on a computer diskette may do so by writing to the Market Bookshop, Depot Ave., Falmouth, MA 02540.

Woods Hole, Massachusetts
August 1986

Acknowledgments

THIS BOOK has been written on weekends, between watches, and is a private view of the sea. It does not reflect the entire body of modern theoretical knowledge about the ocean. It is an account of my own involvement, with others, in a particular aspect of the circulation of the upper ocean over the past ten years.

I would like to thank Wolfgang Krauss for inviting me to Kiel, Germany, in 1976, where the whole adventure began; Fritz Schott for his enthusiastic collaboration in the original beta-spiral study of the historical data; and Eric Firing, Dave Behringer, and Larry Armi for sharing the work at sea during 1978-1981 to get a better data base, and sharing the labor of interpreting it. Bill Jenkins's analysis of the tritium/helium water samples contributed greatly to our confidence in the interpretations. Curtis Collins was a patient and encouraging program officer at the National Science Foundation, which made it possible for us to make five surveys of the beta-triangle area in 1978-1981.

I must also acknowledge the stimulation received by a model worked out by Peter Rhines and Bill Young, which forced me to seek a way to break out of its (to me unrealistic) restrictions, leading to a model with subduction at outcropping of density interfaces in 1983. I could not have formulated that model correctly, or worked out its implications, without the collaboration of Jim Luyten and Joe Pedlosky. It is always an inspiration to see how Joe weaves a closely reasoned mathematical argument. Arnold Arons, Mary Fassett, and Nelson Hogg helped me clarify various parts of the text.

To get beyond the formal model to something more flexible, and to incorporate buoyancy driving, required a new approach. Jim Luyten suddenly saw, one day at the blackboard, that we could formulate a series of problems by the method of characteristics and get solutions by microcomputer. A flurry of activity ensued. Some of the results are described in the latter portion of the book.

My recent enthusiasm, along with Jim's, over the prospects of a living atlas, with oceanographic data displayed on the color screen of a microcomputer, has roots going back to the early 1960s, when I was exposed to the marvelous and exciting intellect of Ed Fredkin, to whom I owe many a happy debt.

ACKNOWLEDGMENTS

Science is both an individual and a social activity. This book may be individual, but the ideas and results, the data and the theories, the work at sea and the funding, have resulted from a happy free exchange among colleagues.

It is also a pleasure to acknowledge the encouragement and help that I received from Alice Calaprice and Edward Tenner of Princeton University Press.

As oceanographic institutions grow larger and more impersonal, there is a tendency for the links to weaken between us academics and the officers and crews who man our research vessels. We spend more and more time in our laboratories and offices. They spend longer periods at sea. Friendships forty years old slowly wither. And yet the data gathered from the ocean is the very lifeblood of our science. I picked a fictitious Chief for the dialogue in this book. It could have been any one of a number of real seamen I have known and respected. No book on oceanography can fail to acknowledge the contribution to scientific knowledge made by the mariners and marine technicians who work our research ships.

A VIEW OF THE SEA

One / Gravity and Pressure

W HEN THE Chief returns to the mess after a brief inspection of the
auxiliaries in the engine room, we resume our conversation. He
is three years older than I. After the Armistice in 1918, when his father returned from the trenches with a severe case of what used to be
called shell shock, his family had a pretty tough time. His first job was
as a lumper, unloading fish at the New Bedford wharves. "I didn't
want to spend my whole life doing that, and I tried to figure out how I
could better myself. I knew that I wasn't smart enough to be a lawyer,
and I didn't have enough money to become a doctor. Then I got a
chance to go to the Merchant Marine Academy and went into the en-
gineering course. During the war I served mostly in tankers. Once you
get used to the heat of the engine room, and the noise, it's not so bad a
life. You get to see the world, you don't get pushed around, and once
you make chief, the sense of responsibility for the ship takes ahold of
you. It's boring—you drink too much—but the days pass comfortably
enough. When I retired a few years ago I just couldn't stand being
cooped up in the house. I'd find myself going down to the fire station
for a game of pinochle. After awhile I guess I kind of wore out my welcome. Even in a fire station the days don't pass so quickly as in a ship
at sea. That's why I take these relief jobs. Just to get away from the
house, the streets, and all the people—back to the life at sea. How did
you get into oceanography?"

I tell the Chief that Pearl Harbor had caught me just as I was beginning a graduate course in astronomy at Yale and finding the mathematics beyond my depth of understanding. I was also caught in a di-
lemma: my pacifist upbringing forbade me to take an aggressive part
in the war. For a time I taught navigation to Navy V-12 students, and
then the astrophysicist Lyman Spitzer found a job for me at the Woods
Hole Oceanographic Institution in research aimed at destroying sub-
marines. That seemed more justifiable to my conscience than bomb-
ing civilian populations, but I have never felt easy about it. So when
the war was over, I had already become so interested in the ocean that
I decided to stay on—with a minimum of formal education. Most of the
other wartime scientists left for other jobs, and the funding for oceanography was uncertain. But I felt sure that there was a lot to find out

about the ocean, and within a few years I began to discover things—
mostly about the ocean circulation. It has been a wonderfully exciting
career for me.

My own understanding of the ocean is based on a construction of
rather simple ideas. I thought that, to make myself understood by the
Chief, I would begin by explaining hydrostatics. So I got some of those
large sheets of paper that computers spew forth, and a No. 2 pencil,
and made the following sketch (Figure 1.1). Let's imagine that the
earth is a perfect sphere. It is not rotating, and it is covered with a thin
layer of water. Gravity is pulling the water toward the center of the
earth, but the water does not accelerate in that direction because it is
held up by the solid bottom of the ocean. It presses down against the
bottom, and each successive layer above the bottom presses down
upon the ones below. Only the very top layer of water is not pressed
down because there is nothing (except the atmosphere, which, as oceanographers, we will ignore) on top of it. You can see that the pressure
increases with depth. The pressure downward on the top of a layer is
exceeded by the pressure upward on the bottom of a layer—the difference being equal to the downward gravitational attraction of the
earth on the water within the layer. This balance of forces between
the vertical difference (the *gradient*) of pressure and the downward
gravity is exact. The net result is that when the forces acting on a
layer are added up there is no net force acting downward (or upward).

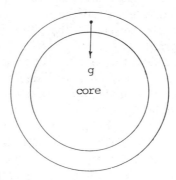

FIGURE 1.1. A section through the center of a nonrotating earth, showing the solid
central core, and, surrounding it, an ocean of uniform depth. The gravitational
force *g* pulls the particles of ocean water toward the center of the earth. The ocean
water does not fall toward the center of the earth because its centripetal acceleration is balanced by an upward reactive pressure force from the solid bottom, or core.

And hence by Newton's laws of motion, the layer is not accelerated vertically. It just stands still on the earth, in a state of *hydrostatic equilibrium*.

You will notice that I speak of layers, whereas the water itself is a continuous substance. Speaking of layers is a kind of fiction, something that we just dream up in our minds to let us talk about the ocean water as though it has distinct parts. I am going to use the idea of distinguishable layers all the time.

The Chief asks, "You mean the difference between a carvel-built and a clinker-built small boat hull?" Well, I reply, yes and no. It is true that in a certain sense there really are layers in the ocean—more or less distinguishable layers of water with differing density, temperature, salinity, and other properties. Looking at drawings of data plotted on vertical sections across ocean basins, you can see that there certainly are more or less horizontal strips of water with visibly different properties—much like the visual impression we get from the clinched planks of a clinker-built hull. But when you look at them in more detail, these strips blend into one another; of course so do the seams between the planks in a poorly focused photograph.

When we introduce our minds to the idea of a hydrostatic ocean, it is probably a good idea to remember that the ocean's water is denser at depth than near the surface. This is mostly a result of compression under pressure, but an important part of the density contrast is due to variations of temperature and salinity with depth and geographical location. Warm water floats on top of cold water, and fresh water floats on saltier water. If we imagine that we have such density contrasts in the ocean, the state of hydrostatic equilibrium on an earth that is not rotating consists of a series of concentric layers; the density increases downward, and the surfaces that bound each layer are perfect spheres. I then sketch another picture (Figure 1.2) for the Chief to show this concentric set of spheres that bound the layers into thin spherical shells of ocean water, with density increasing downward. The Chief looks at it for some minutes and then comments that I had certainly "made the oceans very deep—extending half way down to the center of the earth." He is right: compared to the distance to the center of the earth, the oceans' depth is actually minuscule—scarcely one-thousandth of the distance. I tell him that I drew it that way so that we could see the layers; otherwise they would all be drawn around the rim within the thickness of a pencil line, and an HB6 pencil at that! So every time I sketch something about the ocean, I am going to have to exaggerate the vertical scale of my drawing in order to show anything at all.

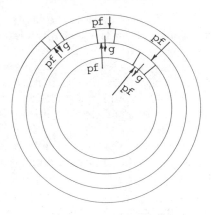

FIGURE 1.2. Here the ocean is divided into three superposed layers, each of which is attracted by centrally directed gravitation. The circles represent interfacial boundaries between the layers, perpendicular to which the internal pressure field exerts pressure forces. The gravitational force g acts in each layer, and the pressure forces, *pf*, act on the layers at the interfacial boundaries. The force due to pressure acting downward on a layer's top surface is less than that of pressure acting upward on a layer's bottom. Pressure itself is one of those quantities, like temperature, that do not have a direction, through it can act perpendicularly to any surface to produce a force. At any particular interface the upward pressure force is equal to the downward pressure force. But when a layer separates two different interfaces, in hydrostatic equilibrium the three forces acting on each layer must balance, which means that the pressure force acting upward at the bottom interface of each layer must exceed that acting downward on the top interface by the total gravitational force acting downward within the layer. Therefore, the pressure itself must increase with depth.

One of the really important things about the great difference between the vertical and the horizontal scales of the ocean is that vertical motions are much smaller than horizontal ones, as far as ocean circulation is concerned. This means that vertical accelerations are generally so small that they can be ignored, and the pressure in the ocean is always computed hydrostatically. On the other hand, the horizontal velocities are larger, and often their accelerations are too. And since they are not locked into the tight embrace of gravity acting vertically (this is our definition of "vertical"—the direction of gravity), we must take the horizontal accelerations into account.

One way to get a horizontal acceleration is by means of a horizontal pressure gradient. It could come about in the following way. Let's draw a small portion of ocean with the earth's curvature straightened out (Figure 1.3). We are looking at a side view of an ideal ocean, on a

flat bottom, in which the water has a homogeneous density. The hydrostatic pressure starts as zero at the surface and increases linearly
with depth. If the top surface of this ideal ocean is horizontal, then the
hydrostatic presusre (top of Figure 1.4) will be independent of the horizontal position—that is, despite the large vertical gradient in pressure, there will be no horizontal gradient of pressure. Pressure, you
will remember, is one of those properties that has no direction—unlike gravity, which is directed toward the center of the earth. It is only
a difference of pressure, or a pressure gradient, that can exert a force.
"I know," interrupts the Chief, "they used to tell us at the Academy
that pressure never burst a boiler, only the difference of pressure inside and outside." Now, if the horizontal gradients of pressure are
zero, and there are no other forces, then the water in this ocean will
not be accelerated horizontally. If it starts at rest, it remains at rest.

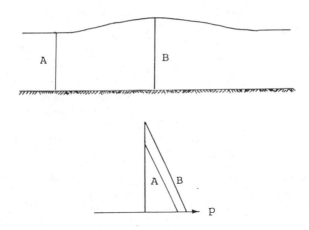

FIGURE 1.3. In the top panel we see a vertical section of a layer of fluid, whose free
top surface has a bump. The solid bottom is shown by a hatched line. The hydrostatic pressure starts at zero (or atmospheric pressure) at the top surface and increases downward, as shown in the graph below. Because the pressure p at the vertical line B starts at a higher level than it does at the vertical A, it is greater at all
depths. The two sloping lines, A and B, show the pressure at depth at the two different verticals. Thus at any given level p at B, under the bump, always exceeds
that at A.

Although we have considered the vertical balance of forces statically, it is ob-
vious that, in the absence of other forces, the bump cannot remain in equilibrium
because the horizontal component of the pressure gradient will accelerate columns
of water horizontally; for example, between verticals A and B, columns will accelerate toward the left.

7

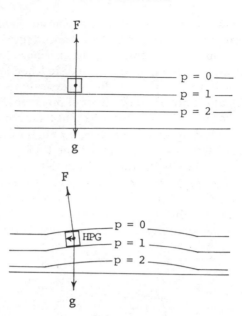

FIGURE 1.4. This figure shows another way of thinking about how hydrostatic pressure in the vertical leads to horizontal forces on water. The top panel shows three levels of increasing pressure, $p = 0, p = 1, p = 2$, increasing hydrostatically downward. Consider the box in the upper layer. It is pushed upward by the excess of pressure $p = 1$ on the bottom surface, denoted by the force F. It is also pulled downward by the force of gravitational attraction g. The vertical forces balance exactly. The forces due to pressure acting on the vertical sides of the box are both proportional to $p = 1/2$ but oppose each other. Therefore, no net forces are acting on the box and it can remain at rest.

The lower panel shows a bump in the surface. The surfaces of equal hydrostatic pressure parallel the curve of the bump. When we examine the force balance, we find that the box is pressed upward at an angle to the vertical. Only the vertical component of this inclined F is balanced by gravitation g. Thus there is a horizontal resultant left unbalanced, HPG. Unless we introduce some other horizontal force to balance it, the block of water will accelerate toward the left. Water on the right-hand side of the bump will accelerate outward from the bump to the right. The bump will therefore tend to flatten out as water moves away from it. In the ocean, bumps associated with large-scale features of the oceanic circulation are so small vertically compared to their geographical width that the vertical accelerations are always smaller than the horizontal accelerations. Therefore, there is no contradiction in assuming that the force balance in the vertical is approximately hydrostatic, but that accelerations must be considered in the horizontal.

Or it could be moving at constant horizontal velocity without horizontal acceleration.

The lower portion of Figure 1.4 shows the same ocean, but this time with a raised bump in the surface. A sudden rainstorm has perhaps just dumped an inch or two of water on the surface of the ocean, over an area with a horizontal extent several times that of the ocean's depth. The vertical balance of forces is still hydrostatic, but because the sea level from which calculation of pressure starts downward is an inch or so higher in the middle of the bump than at the edges, the pressure on all horizontal planes is greater underneath the bump than at its edges. (This can be modified in the case of an imaginary ocean in which the density varies with depth.) So you can see that there are horizontal gradients of pressure. They do not amount to much—a matter of a few inches of water per score of miles, perhaps—but, on the other hand, we have not provided for anything that will oppose them. They are independent of depth, and therefore act similarly at all depths. The result is that they will inexorably accelerate the water, as vertical columns without shear, away from the central high-pressure region under the bump. As long as the pressure on a horizontal plane under the bump exceeds the pressure outside, the outward velocities keep increasing. So when the pressure excess under the bump reaches zero, the water is flowing outward at its greatest velocity, and a low pressure "dent" begins to form where the bump once was. As soon as the low central pressure appears, the outgoing velocities begin to decelerate. And the process tends to operate in reverse. In other words, the system overshoots, and a system of outgoing waves develops. Waves very much like these are known to propagate over the ocean's surface when submarine earthquakes suddenly elevate a portion of the ocean bottom, forcing a bump in the sea surface. These are the waves known as *tsunamis*.

The Chief appears to become a little restless. "Aren't you sort of wandering from the topic of ocean circulation? I thought you'd be telling me about things like the great ocean currents driven by the winds and by heating and cooling—maybe even the Gulf Stream—but not earthquakes."

Well, there are some preliminary ideas about how fluids get moved around that you have to get firmly in mind first, and I hate to have to tell you that we have not finished with them yet.

In the sketches I drew (Figures 1.3 and 1.4) the water was supposed to be of the same density. Now I want you to imagine that there are three layers of water of different density. The interfaces that bound these layers in Figure 1.5 are not drawn horizontally. Nor is the top

9

surface. We can now proceed to calculate the pressure field from the vertical hydrostatic equilibrium. Pressure starts at zero (we don't have an atmosphere here) at the top surface. It increases linearly with depth until we encounter the interface bounding the top and next-down layer. The linear rate at which pressure increases with depth hydrostatically depends on the density itself, so once we cross the interface into a deeper layer the pressure increases more sharply. In Figure 1.5 I have put a bump at the top, where the free surface is, and I have drawn the two interfaces (shown by solid lines) dipping downward, so that there is a central pool of low-density water. This means that although the pressure at a particular horizontal is higher in the center within the top layer, its rate of increase with depth can fall behind at points nearer the edges. This is because we cross the interfaces first, at shallower depths, and the horizontal gradient of pressure can decrease in steps as we descend from the top to the bottom layer. The horizontal arrows in Figure 1.5 depict the size and direction of the horizontal gradients of pressure that might be expected from the ar-

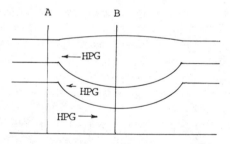

FIGURE 1.5. A vertical section through three superposed layers of differing density (here due to temperature). The hot, upper-layer water accumulates in a dome, with a bump at the surface. Therefore an unbalanced horizontal pressure gradient (HPG) is in this hot layer. The interface separating hot from warm water slopes downward into a dent. When we calculate the pressure in the center of the warm dome at the vertical B, we find that it does not increase as rapidly with depth as it does along the vertical A due to the shorter path through less dense water along the vertical A. This means that the HPG in the warm layer is less than that in the hot layer. As we now pass down into the cold bottom layer we pass through another density interface that has a downward dent. Therefore the HPG in the cold layer is even less than that in the warm layer, and indeed may even be reversed in direction, as indicated in the figure. In oceanography we generally cannot measure the slope of the top surface, so we only calculate the differences between the HPGs in the different layers. This introduces an ambiguity in computing the field of the HPG in actual oceanographic calculations from observed data of the density field.

rangement of the surface and interfaces shown, and from the application of hydrostatic equilibrium in the vertical. The differences in density between different layers in the ocean are generally only a few thousandths of the full value of the density of the ocean water. Therefore, if the surface has an upward bump of a few inches (let's talk like scientists and say 10 centimeters) at sea level, then in order for the horizontal gradients of pressure to be reduced to zero in the deeper layers, the density interfaces between the layers must dip downward under the bump some 100 meters. The changes in thickness of density layers is easily measured in the ocean using thermometers and salinity samplers lowered over the side of a research vessel, but there is as yet no way to determine on shipboard the much smaller elevations and depressions of sea level at the surface. Someday it may be possible to do so by precise satellite altimetry. For this reason oceanographers usually can only talk about the horizontal pressure field at one level relative to another, but not about the absolute pressure field, which would require information on the unknown sea-surface level.

"So far you've only told me about pressure gradients operating in the horizontal. What other forces are there?" asks the Chief. So I explain that there are also viscous stresses. The regular viscosity of water is small and unlikely to be of direct importance in the ocean. On the other hand, turbulent irregular motions of scales between a centimeter vertically and 10 kilometers horizontally might contribute important horizontal stresses—but the trouble with anything related to turbulence is that no one quite understands it. So I try to avoid models that depend heavily on viscosity, in the hope of finding areas of understanding where it does not dominate things. In fact, frictional forces do not seem to be very important in many aspects of the dynamics of the deep ocean.

The reason I am emphasizing horizontal pressure gradients now is because when we turn our attention to the effect of the earth's rotation, we are going to find out that they can be balanced by another kind of force—the *Coriolis force*—and that it is the balance between horizontal pressure gradient and Coriolis force that dominates ocean circulation.

THE CLOCK on the bulkhead of the mess reads 10:30 A.M., and we can hear a noisy automobile speeding its way up the quay, and then quick footsteps on the gangway. A sallow, mustachioed man comes in and asks for the Chief. He is one of the shipping agent's runners. He is typical of that unhappy breed. Used to the often crude abuse of ship stewards and the unexpected demands of ships officers of all nationalities

11

and under the heel of his own boss, he seems half-apologetic—obsequious even—and in a hurry. "We've arranged for a place at the fuel dock tomorrow morning at eight; how much will you want?"—"I think 70,000 gallons of Number 2 will be about right, and we might top off our fresh-water tanks, too," replies the Chief. "Can you come down with me to the office now, to sign the papers? Then we can have ourselves a good lunch." All engineers, from long before the time of Colin Glencannon, are aware of the possibilities of such a ceremony—after all, $100,000 was about to change hands in a dingy agent's office. The Chief fetches his cap, and as he goes out the passageway he calls back, "What I want to know is why the rotation of the earth doesn't throw all the water off into space."

Two / The Earth's Rotation

THE AGENT'S runner brings the Chief back to the ship sometime in the mid-afternoon, but I don't see him until about five o'clock. Then he appears in his shore-going clothes and proposes that we walk into town for a drink and supper. Las Palmas is on a peninsula jutting out from the north end of Grand Canary Island. At the tip of the peninsula is the smaller and rugged La Isleta. The port is on the eastern side of the peninsula, and a more residential part, where we were headed, is on the west along the Playa de las Canteras. Our ship is tied up pretty far north of town, and our route is, at first, something of an obstacle course. We dodge between stacks of containers, avoiding places where the dock pavement had caved in and where scattered piles of broken crates and other rubbish abounds. After we pass the fuel-tank farm and the Muello Pesquero, we wander into a bazaar-type public park, Santa Catalina, which is crowded with stalls offering cheap scarves, flags, trinkets, sandals, shoddy cameras, vile colored ices, and those things that a sailor craves much more. Sailors from Poland, Holland, and Greece and Spanish soldiers are milling around, seeking enjoyment. The narrow streets are noisy with sputtering little cars, and undernourished, unattended children scamper underfoot. When one has seen one of these places, one has seen them all. We press on to the western side of town, where we find a relatively tranquil cafe, above the beach.

It has been a long, dry walk and it is time for a beer. "We've got to be at the fuel dock at eight tomorrow, and I haven't found the Captain yet." The Chief lights a cigarette and takes off his cap. "You gonna tell me more about how the ocean works? And maybe why the water doesn't fly off at the equator?"

IT IS A good question, I think. The complications caused by the earth's rotation are at the heart of understanding the ocean's circulation, and they are somewhat difficult to explain, much less understand. The physical laws of motion, invented by Newton, tell us that particles of mass (or, if we like, the individual droplets of water in the ocean) will remain at rest or maintain a constant velocity along a straight line indefinitely unless acted upon by a force. But particles on a rotating

13

earth, even if at rest with respect to it, do not move along straight lines. They are moving around the axis of the earth in circles, once a day.

"Like satellites?" asks the Chief. No, I reply, and I grab a paper napkin. On it I draw a picture of the earth and the paths that all the water droplets in the ocean (at apparent rest around us) traverse each day, as well as the orbit of a satellite circling the earth just above the atmosphere oblique to the axis. Satellites go around the earth in orbits, the planes of which pass through the center of the earth; but the objects on the earth—the stones on this beach, that tree, you and I, this beer bottle, and our droplets of ocean water—are going around in latitude circles. We are not circling the center of the earth, but around a point on the axis of the earth's rotation somewhere between the poles. But even more important, a satellite in low orbit goes around the earth, in absolute space, much more quickly than once a day. The satellite has only one force acting on it—the earth's gravitation directing it toward the center. This force all by itself is so strong that the centripetal acceleration is also large, as is the rate of revolution in orbit. The particles that make up terrestrial particles are acted upon by at least one other force in addition to central gravitation: they are acted upon by reactions of other particles on the earth. True, a bottle on a table is pulled toward the earth's center by gravitation, but it is also acted upon by an upward, reactive force of the table. The difference between the two forces is small but just enough to allow the bottle to accelerate around the earth's axis once a day. Particles of water in the ocean are subject to gravitation, but they are also subject to pressure forces from the surrounding particles crowding about them. The hydrostatic increase of pressure with depth assures that there is an excess of pressure acting on their bottoms, tending to force them upward—against gravitation. Again, this is not a perfect balance—the out-of-balance is just sufficient to ensure that the particles rotate about the earth's axis once a day. Why don't the particles at the equator fly off into space? They are going around the axis much too slowly, even though it is at a speed of about 400 meters a second. What we do notice at the equator, as a result of the centripetal acceleration, is that gravity is apparently reduced a small amount, by the amount necessary to produce the acceleration required for a circular orbit. The hydrostatic increase of pressure with depth is slightly reduced. And we, so to speak, "absorb" this change of apparent gravitational attraction into a combined quantity that we call gravity. Thus particles do not fly off at the equator. Gravity just seems to be slightly weaker there. This is the kind of thing that can be verified by pendulum clocks. In the eighteenth century measurement of apparent gravity provided a splendid field of study for adventurous physicists who liked to travel.

"Well, what about the particles that don't happen to be on the equator? I'd think they'd slide toward the equator, since gravitation can't pull them toward the center of the circles that they go around," declares the Chief. I tell him that's just what they would do at first! And once they start to do that they pile up near the equator until they build up a slope of the ocean's surface that slows them down. By forming a bulge of matter at the equator, the water builds up a poleward-directed pressure gradient that, in combination with the earth's central gravitation, provides just the right force to maintain the steady acceleration needed to go around in latitude circles. So we see that an earth that is bulged out at the equator can rotate uniformly, and all the particles of which it is made can get the net force they need to accelerate around in circles from a combination of the gravitational force and the pressure gradient. The oblateness of the earth accounts for the various lengths of the degrees at different latitudes. This phenomenon also provided eighteenth-century physicists and astronomers with many reasons for geographical exploration and adventure.

The picture we want to keep in mind is one of a rotating globe whose particles accelerate circularly around its axis and maintain a delicate equilibrium, but not balance, between the pressure field and gravitational force. For a person in a ship, the plane of the horizon defines the horizontal, and the local perpendicular to this plane is the vertical. There is an apparent gravity along this vertical, and we think of it as pointing downward. It does not point directly at the center of the earth, and it is not the pure gravitational force. It includes a small additional effect due to our inexorable circular acceleration around the axis of the earth.

Sensing some mental struggle on the Chief's part, I ask him if I am losing him yet. "Hell, no," he replies, "but my head is going around in circles." We could not have gone on much longer anyway, because the bosun and a couple of the hands have discovered our hiding place, pull up chairs, and turn the conversation to more profane matters.

WHEN I wake up the next morning I discover we are already at the fuel dock. The bosun is on deck helping the fueling gang drag their great hoses across the deck. He looks a little bleary-eyed from the night before but is cheerful enough. When I go down for breakfast, I do not find the Captain at our table, just the Chief. He is drinking black coffee. I get my usual scrambled eggs, bacon, pancakes, syrup, orange juice, and coffee. "The mate moved the ship."—"Where's the

Old Man?"—"He didn't come back from the Campo de Golf—must have met somebody nice."

The Chief is drawing diagrams of the rotating earth on a scrap of paper. "Let's see if I've got this straight. Out in space a big hunk of dispersed matter is pulled together into a ball by its gravitational attraction until the pressure inside builds up big enough to bring the inward motion to a stop. The gravity force would point directly toward the center of the ball, and the pressure gradient that opposes it would point straight out from the center. They'd be in the same line but opposite and equal. Everything would stand still, or at most the whole ball would be moving steadily in some fixed direction. That's what I understand to be your simplest nonrotating case—the one you called hydrostatic." I nod my assent.

The Chief continues: "If the particles that make up the ball happened to be spinning around before they got together by gravitation, then when they clumped together and knocked each other about they'd end up all going around some axis in a kind of solid rotation— by that I mean each part of the ball would go around the axis once in the same amount of time. Now the balance of forces that we pictured in the nonrotating case has to be modified. We all know that to get a stone at the end of a string to swing around in a circle you have to pull on the string with a certain force. That accelerates the stone toward the center and makes it go around the circle—it wouldn't go otherwise. You could rig up something instead of the string if you wanted to—a steam jet pointing inward could push it in toward the center, for instance. Now, that gravitating ball that's spinning is made up of a lot of stones or water droplets that won't go around the axis of the spinning earth unless there is something that pushes or pulls them toward the axis of rotation. Gravity pulls inward but not toward the axis, just more or less toward the center. So here's where the pressure gradient comes in. The surfaces of constant pressure are no longer perfect spheres, as they were in the first case, but oblate ellipsoids—bulged toward the equator. The pressure gradient is therefore not completely balanced by gravity, and there's a small part left over that is not balanced. This unbalanced part just happens to be directed toward the earth's axis, and just strong enough to provide the inward push that every single particle on the rotating ball needs to move around in latitude circles once a day. If it weren't so, the whole ball of matter would not be in 'solid rotation.' It might even collapse or break apart. Here, I sketched it (Figure 2.1). Have I got it right?"

In admiration, I tell the Chief that he has said it better than I did.

16

And he said, "I still find it hard to believe that any mass the size of the earth could settle down to such a perfect equilibrium."

This is a perfect opening for me. Now I can say, well, it is not quite perfect equilibrium, and that is what the theory of atmospheric circulation and ocean circulation is all about. The heat coming in from the sun is what messes up the perfect equilibrium. This heat comes in mostly in the tropics and expands the water and air there by heating them up. Then they become less dense than at higher latitudes, and the pressure gradients in the air and sea that would otherwise be equilibrium with the centripetral acceleration of uniformly rotating air and water are disturbed.

Let me draw what the state of affairs might look like if the top layer of water in a certain band of latitude were heated up. Because each of our layers is thought of as having uniform density, the only way we can have an excess of heat somewhere is to think of having a locally thicker warm layer, as we saw in Figure 1.5. This has been drawn on the sphere in Figure 2.2. Thinking of oceanic scales, we might consider a mid-latitude, say 45°N, and assume that the interface represents the main thermocline of the ocean. Locally it is 500 meters

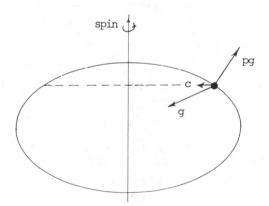

FIGURE 2.1. A force diagram showing how in an oblate rotating self-gravitating globe of water the forces of gravity g and the total pressure gradient pg do not line up and cancel. Instead, they leave a small resultant force c directed toward the axis of rotation that is capable of accelerating the water particle around the axis on a small circle (of latitude) at the angular speed of the globe's rotation. For solid rotation of the globe, this kind of balance must occur everywhere. It is indeed marvelous that it is possible.

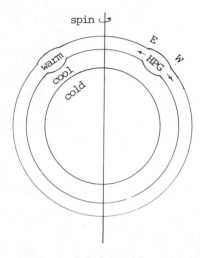

FIGURE 2.2. Density layers in solid rotation around an axis in a self-gravitating mass of fluid. There is a small zonal (constant latitude) ring around the axis where the upper hot layer is thicker. This results in a slightly higher pressure and HPG locally within this layer that are not in equilibrium with the general overall rotation of the sphere. If the water in the northern edge of the ring moves slightly more rapidly around the axis, then the axial component of this HPG can accelerate the water in the ring steadily around the earth at a slightly higher angular velocity than that of the surrounding water. Similarly, the equatorward-directed HPG, at the southern edge of the ring, can be in rotational equilibrium with its centripetal acceleration since it moves slightly less quickly around the earth than the general speed of rotation. For an observer moving with the main rotation of the rest of the globe, these differences of angular velocity within the ring will appear as eastward water currents (E) at the northern edge of the bump, and westward currents (W) along the southern edge of the bump.

deeper than at other latitudes. If the top surface of the sea is 0.5 meter higher at this latitude than at other latitudes, then, with a density contrast of one-thousandth across the interface, the horizontal pressure gradients in the upper layer will point northward and southward from its center, and there will be no horizontal pressure gradients in the lower layers (they are canceled out because of the opposing slopes of the free surface and the interface). We would then have, within this layer, a ridge of high pressure running around the world in a band. The pressure gradient toward the earth's axis would be increased on the poleward flank of this ridge, and reduced on the equatorward side. The delicate imbalance that we had before would be disturbed. On the other hand, if we were to let the water on the poleward flank move a little faster toward the east, it would need this greater pressure gradient to remain in equilibrium (it would need a greater acceleration toward the axis); on the equatorward flank, if we reduced the eastward velocity from that of the undisturbed rotating equilibrium state, it would not need so large a poleward pressure gradient. To an observer on a ship at sea, steaming from the equator toward the pole and crossing this disturbed zone, there would appear to be a westward current on its equatorward flank and an eastward current as the observer crosses the poleward flank. An observer with a thermometer might even notice the higher temperature at depth of the disturbed band as the ship crosses it. The system would still be in a kind of equilibrium, but all the particles of water would not be going around the earth with the same period as the rest of the earth. Moreover, this kind of flow pattern could persist forever, as long as the band of heated water (represented in our layer model as a thickened layer) does not cool off. This might seem surprising, since one would normally expect the band to spread out in latitude until it became of equal thickness. But it is perfectly consistent with the idea of pressure-gravity equilibria in planetary masses spinning on an axis. Not all the matter needs to be rotating with the same period, unless we insist upon a perfectly "solid rotation," which palpably only approximates the case on the real earth.

There is an ocean current that corresponds to an uninterrupted band of lukewarm water moving eastward around the earth: the Antarctic Current. It flows through Drake Passage between South America and Antarctica, and then south of both Africa and Australia-New Zealand. It moves toward the east with a velocity only slightly greater than that of the earth itself. To do so in equilibrium it would need locally a slope upward, toward the equator, slightly greater than that of the equatorial bulge that is already there: about an extra meter across the whole current. From the difference of density across this current we can infer that this difference in "bulge" is actually there. This is an interesting realization of what at first seems like a pure thought experiment. A band of water moves around the earth more quickly toward the east than the earth itself. It needs, therefore, a greater centripetal acceleration toward the earth's axis, and indeed we find a slightly larger than normal bulge in sea surface toward the equator across the current.

What is even more surprising is that the region of an anomalously (warm) thick layer of water does not need to extend all around the earth in a latitudinal band. Suppose, as in Figure 2.3, it is confined to a reasonably small (say, 200 kilometer diameter) circular area in the top layer. We will, for the moment, suppose that it is in the northern hemisphere. Then the eastward current (as seen from an observer moving with the earth) will be on the northern side of the high-pressure region, and the westward current will be on the southern side. Obviously, they are not able to continue around the world in those directions because they would depart from the region of anomalous pressure. We now must inquire whether they can turn to flow in a closed pattern around the high-pressure region as shown. We have not yet discussed north-south flows relative to an observer on the rotating sphere. There is an immediate difficulty. In the equilibrium case, the lower the latitude the faster a particle is moving in absolute space. So if a particle of water wants to move toward the equator it must speed up. Unless there is some force that can accelerate it, it cannot speed up. Here is where the pressure anomaly comes to the rescue. As the eastward-moving particle turns toward the south in our drawing, it gets accelerated eastward (in absolute space) to keep up with the faster-moving (in absolute space) water there. We do not see that acceleration because we are observers moving with the earth itself. So the water seems to move southward on the eastern flank of the high pressure region. Then it flows westward, as I explained before, and then turns northward, where the pressure gradient now conveniently

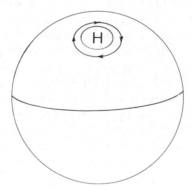

FIGURE 2.3. A small isolated pressure anomaly in a layer of fluid on a uniformly rotating self-gravitating globe cannot support zonal currents around the globe as in Figure 2.2. Instead, the currents turn around the pressure bump and connect to one another, forming a ring current around the pressure anomaly. The text explains the difficulties of making the connecting north-south flows, and that, when the pressure anomaly extends over a large range of latitude, the pressure pattern usually cannot remain stationary.

decelerates it (in absolute space) so that it can join the eastward flow at higher latitude and complete the circuit. In this marvelous way water can be in equilibrium as it flows around an isolated pressure anomaly. And the pressure anomaly does not spread out. The same thing is true in the atmosphere; all weather maps show the circulation around atmospheric high and low pressure regions.

This marvelous equilibrium for fluid particles moving relative to the earth is not generally possible if the gyre around which the circulation occurs extends over a substantial range of latitude. Then there are only very special forms of flow that can maintain their form without dispersion or movement (usually toward the west) in longitude. To try to pursue these questions further here would lead us into complications that we must forego: into many of the most interesting dynamical problems of meteorology. We must confine our attention here to flow patterns that are stationary in the reference frame of the rotating earth, which excludes most of weather and other changing conditions; but it includes steady ocean currents, which are our primary business here.

I have argued this phenomenon in terms of absolute space. Oceanographers and meteorologists find it convenient to introduce the concept of a *Coriolis force*. This is a fake force, just as a centrifugal force is a fake force. It can be derived rigorously by making a mathematical transformation of the hydrodynamical equations from an absolute to a uniformly rotating reference frame.

THE CHIEF says he will need time to digest this stuff. Meanwhile, some riggers have arrived to remove a broken motor from one of the air-conditioning compressors, and he wants to supervise their labors. They go to work with their chain hoists, easing the heavy motor from out of the cramped space in which it was installed. Its replacement had just cleared customs at the Aeropuerto de Gando, and we are mighty happy that it did when we later steam into some sultry weather.

Three / The Coriolis Force

I DO NOT have much time to talk with the Chief for the next few days. My friend Dr. Gregorio Parrilla came down to the ship from Madrid and is taking Bob Stanley and me around for a good tour of the older and more interesting parts of town. Gregorio is a Spanish oceanographer who has done a lot of work around the Straits of Gibraltar. We first met in 1969 when he came along on the *Atlantis II* from Woods Hole to Toulon to work with me in the MEDOC (Mediterranean deep water formation) experiment. Anyone who lives by the sea acknowledges in one way or another the immense maritime heritage of Spain. Gregorio has always been ready to help me when I get lost in the maze of the Spanish literature on marine exploration. He is a splendid host, and we spend a happy day with his family on a little farm back in the hills near the Caldera de Bandama. The countryside is hilly and dry, like southern California beyond the reach of the water mains. The farmhouse is very simple. Grapevines cling to the hills around it, and out back are the remains of an immense old wine press. These are the real Canary Islands that the sun-bathing tourists in the German-speaking hotels with their brass bands, in the "foreign colonies" near San Agustin on the south of the island, do not see or particularly care about. The British, who were tourists here first, are now mostly gone. It is a contrast perhaps like that between California when it was Spanish and now, after everyone else has moved in. The prostitution of beautiful places is a worldwide phenomenon, including at my own Cape Cod.

We sail away at eleven o'clock in the morning. The trade cumulus is too thick for us to see the peak of Tenerife, the site of some of the earliest scientific studies of the upper atmosphere. The outward-bound *Beagle* had tried to land here in 1832 but was forbidden to enter. It was the time of the first great cholera pandemic, and the health officials maintained a strict quarantine. So Darwin was deprived of an opportunity to visit this hospitable place.

We are headed nearly westward, toward a triangular area of ocean where I am about to make a survey of the density structure in order to try to test some theoretical ideas. I have now been there four times. The region extends from 22° to 32°N latitude and from about 28° to

38°W longitude. It is completely on the high seas, beyond anyone's territorial waters. So we do not have to go through the elaborate and dismal process of getting clearances through some state department as we nowadays must often do since the enactment of the United Nations' Law of the Sea agreement, which has become disastrous for science. It is a region mostly in the Trades, free from really bad weather, uncomplicated by strong eddy currents. It is situated in the eastern part of the great clockwise gyre of the temperate Atlantic, with the Sargasso Sea in its center and the northward-flowing Gulf Stream on its extreme western edge. According to the simplest of the wind-driven theories, it is the place where wind-driven circulation is most effective and straightforward. To me it is a good place to test ideas and theories, and that is what my field program is all about.

I DECIDE I better time my discussions with the Chief so that we would get to the details of the reason for the survey a little before we actually got there. There is not much time. We will arrive at the eastern edge of this area, some 10 degrees square, in two days.

We manage to spend about an hour before supper in my stateroom, with a good stiff gin and bitters and a sheet of computer paper laid out on the wretched toadstool-shaped table that is welded to the deck there. Everyone is a little restless the first day out. Lots of wandering around, everybody awake. It takes awhile before people get tired enough to stay in their sacks, or stand watch. Then a ship with a complement of thirty people will seem to be inhabited by only four or five.

The Chief chews on his pencil, then redraws the diagram we had left off with during our last discussion (Figure 3.1)—the one showing the isolated high-pressure anomaly with the current circulating around it clockwise. He says: "You said something about an easy way to think about the physics of the circulation of water around this hump of high pressure that didn't involve remembering the whole earth's dynamical equilibrium—something about a fake force that gave the right results. I'd be grateful for any simplification."

O.K., I think, here we go, and maybe I am going to get into trouble. In any case, let's start with a more familiar fake force.

WHEN YOU swing a stone on a string around your head, you often think of a "centrifugal force" that pulls it outward and stretches the string (Figure 3.2). Actually, the only force on the stone is the pull of the string on it toward the center. It is not balanced by any other force. That is why the stone goes around in a circle: it is accelerated toward the center by the pull of the string toward your hand. If you were to

23

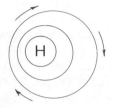

FIGURE 3.1. The Chief's sketch of the flow of currents around a high-pressure region on the earth. The pressure is supposed to be constant along the curved lines: they are called isobars. The current flows parallel to them with the high pressure to the right. Therefore the sense of circulation appears to be clockwise. In the southern hemisphere the sense of circulation around a high-pressure region would be counterclockwise.

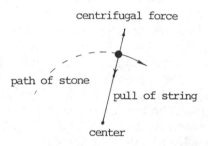

FIGURE 3.2. A stone swung around in a circle by a string is pulled toward the center by the tension in the string. Ignoring gravity, this center-directed pull is the only force exerted on the stone. People often speak of another force—the so-called centrifugal force—that supposedly is needed to make the forces acting on the stone add up to zero. But they do not.

insert a tensiometer in the string, you would be able to measure the tension in the string, and you would know the strength of this centerward pulling force on the string. No centrifugal force acts on the stone to throw it outward. Of course, if you were a bug living on the stone and aware only of the tension in the string, you might think it necessary to have a centrifugal force to balance it. It is a useful, if treacherous, idea. If you want to, you can recast the whole discussion we had about the equilibrium of the rotating self-gravitating globe in terms of a central gravitational force, a pressure gradient, *and* a "centrifugal force" pointing away from the axis of rotation that balances them. Of course, if the forces were really balanced, the particles of the rotating sphere would not go around the axis, but they do so because the centrifugal force is not a real force. So I think it is not really a good idea to think about centrifugal force. But oddly enough, we are going to take the other position when we discuss the Coriolis force.

Now we have to account for the fact that the dynamics of the equilibrium of a rotating self-gravitating globe requires that there be a small (relative to the earth) circulation around an isolated high pressure anomaly (Figure 3.1). Rather than carry in our minds all the huge velocities, accelerations, forces, and pressure gradients involved in the balances in absolute space, we want to isolate the perturbations and interpret them in simple terms that are understandable to an observer who is living and moving with the earth (not out on a stationary space station). We can then show that the perturbations in the accelerations of particles moving around a pressure anomaly can be expressed in the rotating frame of reference of the earth as fake "forces," which we call Coriolis forces (CF).

Let us use this convenient language now to interpret physically the currents going around an anomalously low-pressure center in the northern hemisphere (Figure 3.3) as seen by an observer moving with the earth. I am choosing to discuss first a low-pressure center rather than a high one because it seems closer to our earlier discussion of the centrifugal force.

The horizontal pressure gradient (HPG) is directed toward the center of the low-pressure region. The water is circulating around it in a counterclockwise direction. It is not moving in the direction of the HPG. Why? Because a Coriolis force acts on the water to the right of its direction of motion. This CF just balances the HPG. Because we are talking about such small (relative to the earth) velocities, the CF is much bigger than the "centrifugal force" in this circular motion. Coriolis force per unit mass is the relative velocity times the product of twice the earth's angular frequency (angular frequency is twice pi divided by the period of the earth's rotation) and the sine of the latitude. The curvature of a slow geostrophic flow pattern is not important. So as long as the CF balances the HPG, the flow is along the line of constant pressure. Figure 3.4 shows two other configurations, one with the flow along straight lines, the other around a high-pressure region where the HPG is outward and the CF inward (impossible for a centrifugal force).

The upshot of the introduction of the idea of the Coriolis force balancing horizontal pressure gradients is of practical and theoretical importance in oceanography, for if we can determine the pressure field in the ocean well enough to map it, then we can construct streamlines of flow that are parallel to the isobars, and, depending on the latitude, assign velocities of the current to them. This kind of velocity,

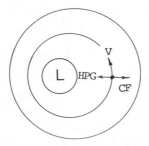

FIGURE 3.3. Counterclockwise geostrophic motion around an isolated low-pressure region. The HPG is directed inward, toward the center of the low. In order to speak of a balance of forces, we introduced the fake (or "virtual") Coriolis force (CF)—a useful fiction that frees us from having to consider the small perturbations of angular acceleration that are actually involved in the absolute reference frame.

which reigns over much of the sea, is called a *geostrophic velocity* (GV). It is measured relative to the rotating earth in the same way that we measure the velocity of our automobiles or of a river flow, neglecting the large extra velocity in absolute space because we are actually riding on a spinning earth hurtling through space.

The Chief twists in his chair and comments that if things get much more complicated, he is going to lose the thread. But he summarizes my explanations nicely: "You mean to say, I take it, that geographically isolated anomalies of the pressure field that would disperse into gravity waves and eventually disappear on a nonrotating self-gravitating globe will not disappear so readily on a rotating globe. In fact, the HPGs associated with them can be balanced by local CFs, and they can persist longer. That is really quite marvelous. What you are telling me now is that this precarious out-of-balance of forces that

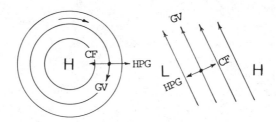

FIGURE 3.4. If the isobars are straight lines, as in the right panel, there is still a CF. In the left panel we show circulation around a high pressure center, with the CF directed inward, toward the center.

keeps all the particles rotating around an axis in a uniformly rotating globe is so stable that when it is perturbed slightly by small buoyancy sources (the radiation from the sun, for example), the perturbation HPGs can maintain a nearly stationary (with respect to the rest of the sphere) pattern, together with a nicely adjusted out-of-balance of the particle accelerations within the perturbed region. Here we are on shipboard, and we're measuring everything with respect to the equator and the Greenwich meridian; we can interpret these relative accelerations as Coriolis forces whose strength is proportional to the current velocity that we can measure. It also means that large lumps of warm water can reside in the midst of colder water, despite the HPGs, without immediately spreading out in waves, as long as they have a ring of current encircling them and holding them in by their associated CFs. You know, I kind of like Coriolis forces. Maybe you ought to grant them the dignity of being called 'virtual forces' instead of 'fake forces.' "

I decide to elaborate a little more, using familiar territory. My home is on Cape Cod. It is a place that was swept over by a great ice wall two hundred centuries ago. Nothing that man can do could have stopped that implacable glacier at the time. Today there is another great wall lying to the south of us in the ocean. It is a wall of warm water, some 900 meters thick, that bounds the warm-water pool of the subtropical gyre to the south and the cooler shelf water to the north. One would suppose that in the nature of things the warm-water pool would try to spread out over the cool shelf water and reach all the way to Cape Cod. It would improve our winter weather if it did reach us, but it does not. Why?

I sketch the situation for the Chief (Figure 3.5). The block of water is cool (C) to the north near Cape Cod. There is a deep pool of warm water (W) on the southern side. The two water masses are separated by the liquid wall that reaches up to the surface. In order to avoid having HPGs in the deeper, cool, and denser layer under the wall, the sea surface on top of the warm pool must be higher than that on the cool region. I sketch this as shown in the figure. Now, of course, we have a HPG in the warm layer pointing toward Cape Cod. This is why we apprehend an imminent northward flow of warm water. And it would do that very thing except for one more detail. We imagine that there is a current of warm water flowing eastward along the wall, shown by the arrows. this is not a fiction; in fact, it is a crude representation of the eastward-flowing Gulf Stream.

"And now," rejoins the Chief, "let me risk completing the story: It is the Coriolis force acting toward the right of the direction of flow of the

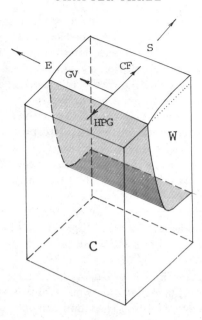

FIGURE 3.5. Block of ocean straddling the Gulf Stream as seen from a point looking southeast from a viewpoint above Cape Cod. The arrows indicate the eastward-moving warm Gulf Stream water. The surface elevation of the warm water is much exaggerated, and provides an HPG directed northward, balanced by the southward CF of the moving stream. The large downward slope of the "wall," or interface separating the warm and cool layers, reduces the HPG in the cool layer under the Stream so that there does not need to be a strong current there.

Gulf Stream that provides the force needed to balance the horizontal pressure gradient in the warm layer, and that is why the wall doesn't surge northward. So the warm water doesn't reach your Cape and goes merrily on its way to Northern Europe, where they need its warming influence much more."

WE DOWN the gin, and go in to supper. As always, the food on Woods Hole vessels is good and the servings are ample.

Four / Measuring the Density Field

A FTER SUPPER I call our first scientific meeting in the ship's library. There are several new technicians who had not been to sea before, and a couple of graduate students. First, I warn them about safety at sea. The watertight doors are heavy, and, if they swing free on a sudden roll, can easily crush a foot or hand. Some bulkhead doors have high sills that can inflict a painful shin bruise. Gear should be firmly tied down on laboratory benches. A little carelessness on the fantail at night can mean falling overboard. It leads to certain death. Only two people have been injured on my cruises. One was a girl who was thrown violently across the main lab of the *Atlantis II* against the projecting steel wheel of a watertight door in the Gulf of Lyons during the mistral, the cold, dry north wind that blows over the Mediterranean coast of France. She had to spend some time in a French naval hospital, though they did not set her wrist right, and it had to be re-broken and reset when she returned home. The other person nearly ground off the end of his finger when he caught it in the clutch of a winch; he later became a distinguished theorist. It is the responsibility of the Chief Scientist not to let such accidents happen, and I wanted to make sure they did not happen again. Not only is it an unhappy event for the victim, but the days lost in getting someone to a hospital can consume enough ship time to wreck a survey.

I also explain our plan of operation for the next three weeks. Taping the map of the area (Figure 4.1) against one of the bookcases, I point out the triangular area of the previous survey. Around the perimeter of the region are marked the positions of thirty stations at which we had lowered an instrument called a CTD for measuring electrical conductivity, temperature, and depth of the water from the surface to the bottom. The device was invented by a gifted engineer named Neil Brown, who has gone into the business of manufacturing this product. I also want to set up some extra stations in the center of the triangle. Our object on this trip is to determine in detail the shape of the density surfaces in the water column in this region from top to bottom.

Somebody asks why the triangle is so big—isn't a lot of steaming time wasted at the expense of getting to more stations? The answer is that we are trying to determine as best we can the depths, slopes, and

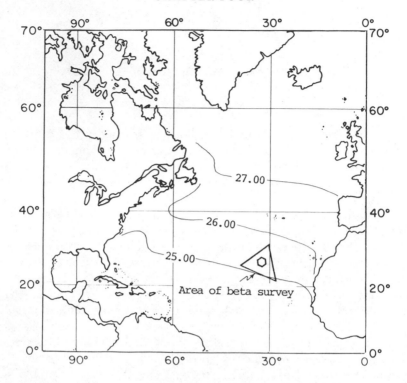

FIGURE 4.1. A chart of the survey plan shown to my scientific team on the night of our conference at sea. The triangle shows the area where CTD stations were taken on an earlier cruise, and which we are about to repeat. The curves that cover the ocean are contour lines of the wintertime density of the water at the surface, expressed in units of sigma-t. This is a shorthand way of writing the density, which never exceeds 1.00 by very much. Thus the sigma-t number 25.00 means a density of 1.025. Wintertime density is greatest at high latitudes, where the interfaces outcrop at the surface. As we lower our instruments we encounter these interfaces again at depth.

curvature of density surfaces associated with the mean circulation of the gyre. But, as the drawing in Figure 4.2 shows, we expect to find a lot of jitter in our plot of the depths we observe because of the eddies that are approximately 60 kilometers long. From our point of view, these transient signals from the eddies are a kind of noise that masks the features we want to see. So we have to space our measurements along the side of a triangle at points far enough apart so that they embrace many of these eddies, and by smoothing the observed depths we can more or less filter them out. Of course, if we made the triangle even larger we would consume an inordinate amount of time getting around, and we also would not really get a local determination on the scale of the gyre. So the area that we are using is a compromise between number of observations, amount of steaming, and the degree of localness on the gyre scale. We have chosen the region in the eastern portion of the gyre to avoid the much more energetic eddies (greater noise) in the western basin of the Atlantic.

We choose our latitude so that we will be working in a region near the boundary of the Trades and the Westerlies, where the southward average transport of the gyre is at a maximum.

Somebody wants to know if she can telephone home. We do have an amateur radio set on board, and can establish a ham-link; But it is really better not to try to communicate with the shore unduly. Generally we do not even hear the news while at sea. A month may pass and we may hear about only one thing, such as Marilyn Monroe's suicide. In a very real sense the rest of the world just ceases to exist. Our loved ones go into a kind of limbo. It's the best way to handle the stress of separation. It is a solace of the sea.

We schedule a test station for tomorrow morning. It is always a good thing to do. Everyone gets a little practice. If there is something wrong with the gear, we have time to straighten the trouble out before we get to the first real station. Our estimated time of arrival (ETA) at the first station is at about midnight tomorrow. The Chief does not say

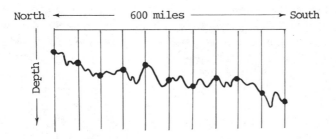

FIGURE 4.2. A side view of one of the sides of the beta-triangle, depth against distance along the section, showing a possible configuration of a single sigma-t contour, and the eleven station positions (the only data is at the level heavy dots) at which its depth is actually measured by the survey ship's CTD. The large-scale trend of the sigma-t curve is reasonably well determined despite the small-scale wiggly nature of the actual instantaneous curve, due to passing eddy turbulence and the higher frequency of internal waves. Clearly, eleven stations are a bare minimum of stations suitable for the task, but a more densely spaced number would consume all the shiptime on a single side of the triangle.

31

anything at the meeting, but I know he will be hanging around the winch in the morning when we make our test lowering of the CTD. Tomorrow will be a long day, so I crawl off to my room and hit the sack.

I wake up while it is still dark, have some coffee in the dim red light of the bridge with the second mate, and watch the sun rise. To me that is the most poignant moment of the day. The horizon reveals itself at the cold, grey edge of the sea; dark cumulus clouds are outlined against a purple, lightening background sky. Rays of light shoot up from the east; the tops of the tallest clouds are successively illuminated scarlet and then gold-yellow. And then the cadmium crimson disk of the sun begins to break through beneath the distant darkness where the sea and sky meet. It is the eternal reaffirmation of warmth and life blazoned across the vista of an implacable, indifferent ocean.

By the time we set up our instruments, make the terminal connections to the cable on the winch, rig the boom, and lay power on the winch, it is 9 A.M. This first test station is bound to be a bit confusing, and of course, as always seems to be the case, the Old Man schedules a boat drill at 11. Just to keep us on our toes, I guess. Or maybe to show that the deck has things to do, too. Anyway, the Chief is there to watch over the winch. The winch and the 7,000 meters of armored electrical conducting cable wound on it are essential for our work. The cable is 5/16 inch in diameter, and when it is fully out, it can sustain a considerable stress. As the ship rolls, the tension varies irregularly and tends to unlay the armor. Violent rolling or a too-rapid descent can throw bights into it. Bights lead to kinks, and kinks to broken insulation or broken conductors and even to total cable failure and loss of the expensive instrumentation—$30,000 to $60,000 gone in one snap. If there is no longer a spare cable, or a spare CTD, it is the end of the cruise, a loss of much time and energy, and maybe $500,000, all told. If someone, in the wee hours, misreads the fathometer, it is quite possible that the CTD will entangle itself on the bottom.

The cable is also under great tension when it is wound up on the winch, when the instrument is raised. If the wire does not lay perfectly level on the drum of the winch, open spaces will develop in the lay, and the increasing irregularity will produce intense pressures at certain points of the cable crushing it. Troubles may develop in the winch itself or in the motor driving it. Under load, the brushes tend to arc on the commutators. Hours of steady laboring can burn them badly. A good engineer is alert for this kind of trouble. If the ship drifts over the cable, damage can occur, even to the ship itself; on one occasion a cable sawed through a ship's hull like a band saw. The lowering and raising of a CTD takes almost two hours. With halts to collect water samples the work can be tedious, and a sleepy winch operator can pull an instrument out of the water, right up to the block at the end of the boom, break off the wire, and lose everything. And the process of putting the whole thing overboard in rough weather and bringing it back involves a moment when it is freely swinging about, threatening to smash into the side of the ship, drop on someone's feet, or pull somebody overboard. It is a deadly pendulum swinging back and forth in arc of about 12 feet just at body height. It weighs about 400 pounds. The possibility of accident to an inexperienced hand is always looming.

So it makes good sense to make one practice station. And then in the middle comes that damned boat drill—timed, it would seem, to cause the maximum confusion.

The CTD instrument has a conductivity cell, a platinum resistance thermometer, and a pressure gauge. Using standard physical tables of the properties of sea water, the computer calculates the temperature, salinity, and density of the sea water as functions of depth and pressure.

There is always some question about how stable the conductivity cell is, so the CTD is fitted into a frame carrying a rosette of sampling bottles that can catch water when triggered from the laboratory computer console. The conductivity of these samples is measured on board in a standard conductivity bridge against standard water. In this way the CTD conductivity measurements are calibrated. The water samples can also be used for various chemical determinations, such as dissolved oxygen, and for radio-isotope tracers such as tritium and its daughter, helium-3.

These water samples are drawn after the instrument is brought aboard after each lowering, lashed down, and the ship is underway to the next station. It is a cold and wet business.

With luck the ship stops for only about 90 minutes at each station, and with the stations spaced about 60 nautical miles apart, as in our triangular pattern, there is a respite in work on deck of about five hours between stations. In the meantime, there is plenty of work to do inside the lab. It goes on in all weather, day and night, 24 hours a day. When there is a big enough scientific party on board, we divide the work into three watches of four hours each. At night the watches seem very long indeed.

The motion of the ship makes every simple operation—punching a keyboard, reading a thermometer, pouring a sample, soldering a connection—a feat of skill. It is tedious, monotonous, physically exhausting. It makes you feel your age.

Everything goes well on our test station, although because of some false starts and rigging problems it takes four hours. With everything aboard and secured, we signal to the bridge that we are ready to go on to the first station on the perimeter of the triangle. If everything went well there, we would have time to make the central stations before heading in to Cadiz. The estimated time of arrival for station no. 1 is midnight, but the weather is beginning to come up with a head sea and we might be a bit late getting there. We are anxious to get started on the long run of stations—maybe as many as forty. Once things got started, they would settle down to a routine. I join the first mate and Captain Robert Munns on the bridge, and we talk about old times and shared experiences: our hurried passage through the Banda Sea during the local civil insurrection, when thousands of Chinese shopkeepers were murdered; the ceremony of the equatorial crossing when I was King Neptune's Royal Baby; and the black night when the most extraordinary, luminous bands appeared mysteriously in the sea.

Five / Geostrophic Velocity

FOUR DAYS have lapsed since we started with the first CTD station on the triangle. We are just finishing the eleventh station, and are about to plot up the density profile along the first side of the triangle, the southwestern side. The Chief has taken some interest in seeing the data come and get plotted by the computer, station by station. The graphs that he seems to like best are the ones that show density against pressure and density against depth beneath the surface. He says they give him "food for hydrostatic thoughts." He had asked me earlier if we had any way of determining the height of the sea surface as we move from station to station, and I had told him that we did not. Sometime ago the satellite people promised to get an accurate sea level of better than a centimeter, but they had not been able to do so yet—for our nonmilitary use, at any rate. He had then asked, "Well, if you can't measure how sea level varies from station to station, how the devil are you going to find out how the pressure varies along various horizontals at depth? Sure, you have the hydrostatic balance in the vertical in each station location to work with, but you don't have a common starting point for all the stations. After all, you start your hydrostatic calculation at the top surface with a pressure equal to zero, or with the atmospheric pressure, and then you calculate downward for each station. This gives you an indication of how pressure varies with depth beneath the sea surface of each station. But to get the HPGs acting between the stations, you need to know how sea level varies from station to station, and this is unknown to you. How do you get around that?"

ONCE MORE the Chief has put his finger on a real difficulty. The traditional answer is that we have to *assume* that we know a depth where the HPG is zero, for each pair of stations, and work the hydrostatics backward to find the HPGs at all the other depths. Since the total amount of mass in each water column above the level of zero HPG has to be the same, by hydrostasy, we can also determine sea level. The difference in sea level never varies more than a meter from one station to another, so that is no problem to us when specifying the vertical distribution of density at each station. We cannot measure the depth to much better than a meter, anyway, because the wire never stretches straight down, and our methods of measuring the actual pressure with pressure gauges are not much better than that, either. So the small variability of sea level does not affect our determination of density versus depth, even though it is essential in determining the pressure.

Let me try to illustrate this with the diagram in Figure 5.1. Here we see two water columns, 1 and 2, side by side, separated by an impenetrable barrier b. The layers of different density are shaded. You will notice that column 2 has thicker and deeper layers of the lighter surface layers. The depths of the first interface down are labeled j_1 and j_2, the depths of the second interface down are labeled k_1 and k_2. The sea levels themselves are labeled i_1 and i_2. Now, it is clear that we can calculate the pressure hydrostatically in each of these water columns, and that, starting at the atmospheric pressure sea surface, it increases less rapidly with depth in column 2 than in column 1. On the other hand, it begins higher up, so we still do not know the pressure difference across the barrier b, unless we know the levels i. Suppose we assume, however, that there is a small leak through the barrier at a depth z. Then, if there is a pressure difference across the barrier, a small flow-through will occur and the sea levels i will change; but the layer thicknesses will not change much because they are much thicker than the difference between the i's. In the case illustrated, the pressure is now equalized between the bottom layers in both columns, and there is still a difference between the pressure in the other layers.

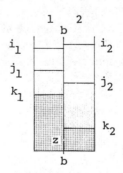

FIGURE 5.1. Two columns of water with different thicknesses of three density layers, separated by the barrier b. As drawn, there is a small leak in the barrier at depth z that equalizes the pressures between the layers at that particular depth—in this case the two bottom layers.

The sea levels i are now determined. There is no discernible difference in the layer thicknesses—less than a meter, essentially unmeasurable.

If we want to, we can place the leak at a different depth—we could put it between the top layers in the two columns. Water would then leak from column 2 into column 1, and the sea level would equalize. The depths of the interfaces j and k would not change measurably.

What I am illustrating here is how one assumption—that there is no pressure difference at a particular level—enables us to compute the pressure differences at other levels. This need for an assumption arises from the fact that we actually do not measure sea level itself. It is not too hard to visualize how in this example the difference of pressure between two columns of water separated by a barrier translates itself into the idea of a horizontal pressure gradient between columns of sea water at different geographical locations. With such an assumption, then, we can calculate the HPGs in the rest of the range of shallower depths.

These horizontal pressure gradients, which we must always qualify by calling *relative* HPGs—that is, relative to the depth where we assume zero HPG—can then be used to calculate the geostrophic velocities (GVs) that will provide just the right Coriolis force (CF) to balance the relative HPGs. To play safe, we had better call these GVs *relative* GVs—also relative to zero GV at the level of zero HPG.

"Of course the relative GVs are going to be different by a constant amount if you choose another reference depth, although I can see that the vertical form of the profile of HPG and GV won't be changed. It's just a shifting of the zero," comments the Chief. "But how do you decide what the level of zero HPG and zero GV is?" The truth is that oceanographers do not have a good way of making that choice. In fact, they have usually argued that the currents in the deep ocean are probably very slow, so we can assume that the HPGs vanish there. The Chief winces: "That may be a way out for a high-paid theorist in a computing center in the Colorado Rockies, but I wouldn't want to bet on it if I was planning a deep dive." So I am happy to be able to explain to the Chief that one of my reasons for studying this area of the ocean is to find a logically and physically consistent way to get the answer to this question.

"It seems to me that somebody would have found a method years ago. What are oceanographers doing, anyway? Spending their time raising money in Washington?" I didn't try to answer that one.

I place the computer-produced graph of HPG and GV (relative to 1,500 meters depth) on the laboratory drafting table. Because the

HPGs calculated along one section all lie within the plane of the plot, they are really only one component of the true HPG. The other component is perpendicular to the paper, and we did not measure in that direction. When we made our first section, we traveled along a straight-line course. Therefore, we can now look only at components of the GVs perpendicular to the plane of the section, but not tangential to it. The perpendicular GVs represent the flow of water across the section. Southwestward flow across this face of the triangle is out of the triangle. When we have surveyed all three sides, we will have an estimate of relative GVs of the amount of flux of water in and out of the triangular prism's volume, for every different kind of density.

I do not think that the small-scale detail is permanent in this region, so I have smoothed out the picture of this relative GV across this southwestern face of the triangle and sketched it in Figure 5.1. It shows weak southwestward flows over the full width (600 miles) of the section in the upper hundreds of meters, amounting to about one centimeter a second (half a mile a day). This velocity falls off gradually with depth, to zero at a depth of 1,500 meters (by assumption). The velocities beneath this reference level are very weak. We do not know the velocity within the plane of the section.

"As I see it," the Chief finally says, "your method of getting at the velocities from the density measurements is only partly successful. You have to assume what it is at one level to find it at other levels. Of course that's better than nothing—but so's a ham sandwich."

He scratches his head and goes on: "The fact that you can't get the velocity components tangent to the sides won't be so bad when you have gone all around the triangle. At least you'll have a measure of the total flow in and out of the triangular region, at all levels and for each density layer. If you know that over a long period of time there isn't any change in the amount of water inside the triangle, then that knowledge of inflow and outflow in each layer ought to be worth something to you. Maybe you could even work out the reference level that you're so concerned about. You know, one of my assistants sounds the fresh-water tanks on this ship every day with a rod. Even with automatic water gauges and our evaporators to make fresh water we don't want any nasty surprises. And that reminds me: you haven't said anything about conservation of water in the ocean yet."

Well, it does seem to be the right time, I confess, so I draw the following pictures (Figure 5.2a-c) on the little blackboard in the Chief's cabin—the one on which he keeps notes about fresh-water and fuel levels. Here we have a layer of constant density in the ocean, and two vertical surfaces, 1 and 2 (hung like curtains) representing surfaces of constant pressure difference. The geostrophic flow, in which the pressure gradient acting from isobar 1 toward isobar 2 is balanced by a Coriolis force, has to be confined within the tube defined by the top and bottom of the layer and the two pressure curtains. It is a kind of tube having a rectangular cross-section. If this tube lies within a layer of water of a particular density inside the ocean, if it is not mixing across the top or bottom, and if water is being neither created nor destroyed (the law of conservation of mass), then the discharge of water

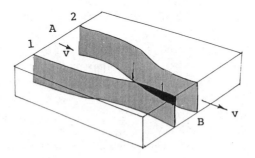

FIGURE 5.2a. The block shows a density layer of uniform thickness. Two curtains, 1 and 2, represent isobars of constant pressure difference. The pressure is higher on 1 at a fixed depth than on 2 at the same depth, so that the HPG is directed from 1 to 2. A geostrophic current flows through the tube formed by these curtains and the top and bottom of the layer. Geostrophic velocity (GV) cannot flow through an isobaric curtain by definition. The GV has to be greater at section B than at section A because the HPG is greater. Mass is conserved.

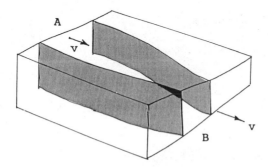

FIGURE 5.2b. This block differs from the previous one in that the layer thickness is not uniform. The isobaric curtains are each of a different height, but the height of each is constant along its length. Mass is still conserved along the stream tube, and the GV is larger at section B than at A.

through all sections of the tube must be the same, even when the tube changes the shape and area of its section. In Figure 5.2a the density layer has a uniform thickness, but the separation of the two pressure curtains varies. This means that where the curtains are close together, as at section B, the geostrophic velocities must increase in order to carry the same discharge. This is quite consistent: when the curtains are closer, the HPG is greater, as is the required CF, and this means the GV will be higher.

Figure 5.2b is a variation in which the density layer is nonuniform in thickness. But the pressure curtains are arranged so that the height (from top to bottom of the layer) of each curtain is constant. Each curtain may have a different constant height, however; for example, here pressure curtain 2 is not as high vertically as curtain 1. The section of the stream tube is not rectangular anymore, but the geostrophic discharge turns out to be independent of the separation of the curtains at any particular section. Up to this point we have been ignoring the fact that the formula by which we calculate the CF depends upon latitude as well as GV. In the two examples we have shown the flowlines of the GV are along contours of constant thickness of the density layers. Contours of constant thickness can be called *isopachs*.

Let us now see what kind of trouble we get into if we let the pressure curtains array themselves across a layer of nonuniform thickness in such a way that their vertical height diminishes in the downstream direction (Figure 5.2c). The GVs would be the same as in Figure 5.2a,

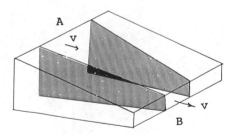

FIGURE 5.2c. Here the layer thickness and isobaric curtains are skewed with respect to each other so that the curtains get shorter in the direction of the GV. This means that with constant latitude the discharge gets smaller as the tube becomes vertically thinner, and mass cannot be conserved. If the latitude at A is much greater than at B, we might be able to maintain mass conservation of a GV field, as explained in the text.

but the discharge in the tube would diminish rapidly downstream from section A to B because the vertical dimension of the tube is shrinking. (Remember that we are still ignoring the latitude dependence of the Coriolis force formula.) So, either we are going to have to violate the idea of conservation of mass in individual density layers (that is, allow fluxes from one layer to another in the interior of the ocean), or we must arrange our pressure curtains so that they always have uniform vertical height, or take the dependence of the formula for Coriolis forces on latitude to rescue us from an unpleasant trilemma.

For example, suppose that there is a large variation in latitude across the block of water shown in Figure 5.2c. Maybe the section of the stream tube at A is at 30°N, and at B it is at 10°N. Then, according to the formula for Coriolis force, you need about three times more GV to balance the same HPG at B than you need at A. To preserve mass conservation, the vertical thickness of the section at B can be only one-third that at A. This can all be subsumed in the simple idea of conservation of a quantity that we can call *potential thickness* (PT). Potential thickness is the local thickness divided by the local sine of the latitude. Suppose that the vertical height of the pressure curtain 1 at section A is 220 meters. Now, the sine of the latitude at section A is 0.5, so the potential thickness there is 440 meters. If potential thickness is conserved as the fluid moves along the curtain 1 toward section B, where the sine of the latitude is approximately 0.17, then the actual thickness there must be about 73 meters, which is consistent with the sketch in Figure 5.3c. I am going to resist the temptation to coin a word to describe lines of equal potential thickness (how do you like isopotpachs?), but I am going to introduce the notation PT to stand for potential thickness. So, evidently, we are going to find that in the open ocean there is a possibility that lines of equal PT coincide with isobars. It is worthwhile trying to find this out, because it is a test of our ideas about the nature of the flow in the deep ocean. The cruise that we are on is aimed at making this test. That's what it's all about.

"Let me be sure that I've got the implications of our discussion of flow through these tubes correctly," interrupted the Chief. "It seems to me that understanding Figure 5.2c is crucial, so let me try to summarize in my own words. We're concerned with the conservation of mass in the tubes. The tubes are bounded on the top and bottom by density interfaces, and on the right and left by vertical surfaces that intersect level surfaces within the layer at two different constant pressures. Within the tubes the geostrophic velocities are independent of depth. Water isn't supposed to flow across the top and bottom interfaces, nor can it flow across the side isobaric walls by definition of geostrophy. So it can only flow along the tubes lengthwise, as from A to B in the figure. By conservation of mass the total discharge through the cross-section at A has to be the same as through that at B. Now we can note that changes in the width of the channel have no effect on the discharge per unit depth, because the difference of pressure across the tube horizontally is independent of the width.

"Now suppose that the cross-section at B is nearer to the equator than that at A. Because geostrophic velocities vary inversely as the sine of latitude, the transport per unit depth of the tube at B will now be greater than that at A. The only way to cancel out this latitude effect is for the vertical height of the tube to be directly proportional to the sine of latitude. Then the two sines cancel out, and mass conservation is assured. With smaller height of tube at B, columns of water must shrink vertically as they move from A to B."

I tell the Chief that he stated the ideas correctly and clearly. I ask him whether we should continue the argument or wait until another time.

The Chief is silent for a few moments, and then he ventures: "I'm going to wait five more days, when you've finished the survey of the second side of the triangle. Then you'll have the depths of density surfaces at three corners of the triangle. We use those depths at the three corners to find the slopes of each density layer and how the lines of constant PT lie on them. Then we try to figure the isobars—that is, the pressure curtains—in each of the layers, and see if we can get a consistent picture in all the layers at once. I'm going to be very surprised if there isn't a catch somewhere. I think it's interesting that by taking mass conservation into account we suddenly discover the useful idea of potential thickness and its urge to conserve itself." With this challenge ringing in my ears, I go back on watch. I am not particularly worried, because I had had four looks at the triangle before, starting in October 1979.

Six / A Chinese Puzzle

ON OUR SECOND leg along the northwestern side of the triangle we head into the Trades, so we are going a little slower, and we pitch more—the first leg had been all rolling. The Chief is hanging around in the lab, watching the girls at the CTD computer console and the people making the salinity determinations on the samples. When off watch, I spend a little time on the bridge with the Mate, sometimes listening to the BBC: the Merchant Navy program; chronicles of strikes and piracy on the West African coast, of armed assaults on shipping in the Persian gulf, and of more vessels being laid up; and the sad news of indications of the continuing decline of the British merchant marine. When we arrive at a station, I slip down to give a hand, and to make sure that matters are going smoothly and safely. Eventually we arrive at the end of the second leg, and we are able to draw up the density structure on two of the faces of the triangle. After much smoothing to get out the eddy noise, and some editing, we produce the straight-line diagram of density layers shown in Figure 6.1. It shows only the top kilometer because the whole picture is somewhat more uncertain at greater depths. The vertical density gradient gets weaker down there and it is harder to discern the mean slopes of the density surface above the noise.

"I can see now that the density interfaces in the region are not all parallel. They slope in slightly different directions. There is a fairly uniform slope down toward the northwest along side 1, but the slopes along side 2 are quite different with depth. They are strongly upward toward the northeast in the upper layers, and nearly flat in deeper waters." So observes the Chief. I had chosen four density surfaces for this schematic sketch, numbering them downward. Though there are five layers, only the top of the lowest layer *e* is shown—it extends beneath one kilometer into the deeper water below. Layer *a* is exposed to the surface, but layers *b*, *c*, and *d* are completely submerged in the triangle.

Now we are going to find out whether this structure is consistent with the physics we have advanced over the past two weeks. Specifically, we want to apply the following rules and see whether they are consistent:

RULE 1. The horizontal pressure gradient, HPG, increases as we ascend through a density interface from one layer of homogeneous density to the one above it by an amount proportional to the product of the density contrast and the interfacial slope, and the increase is in the direction of the upward slope.

RULE 2. Within a layer of uniform HPG, the geostrophic velocity, GV, is also horizontal and independent of depth and is directed to the right of the direction of the HPG (in the northern hemisphere). The amplitude of the GV is proportional to the amplitude of the HPG divided by the sine of the latitude. A corollary is that the change of GV as one ascends across a density interface is to the right of the change in HPG

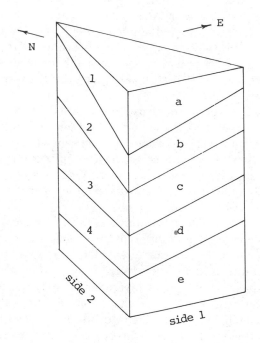

FIGURE 6.1. A perspective view of the volume of water in the triangular survey region. The point on the left points northward. Only the upper kilometer of the water on the vertical sides is shown here. Depths of four density interfaces are drawn schematically in highly smoothed form on the two sides visible to the reader. The interfaces are labeled 1, 2, 3, 4. The layers between them are labeled a, b, c, d. Only the top of layer e is visible, the rest being deeper than one kilometer. As shown, the slopes of the density interfaces are not all in the same direction or the same amount.

and proportional to the amplitude of the change in HPG divided by the sine of latitude.

RULE 3. Because the GV is independent of depth within an individual density layer, the water within a layer moves as a vertical column, without tilting with respect to the local vertical. To conserve mass, any movement in the north-south direction requires that the vertical height of the column vary as the sine of the latitude. The height of the column is simply the thickness of the layer, T. Therefore the GV in the layer must be parallel to contours of a quantity we call potential thickness, PT (where we define PT as T divided by the sine of latitude). Because a column follows a contour of constant PT, its height (T) is proportional to the sine of the latitude.

When we use these rules to study a set of density layers, separated by sloping density interfaces, from real ocean data, we are of course faced with the fact that to begin with we really do not know the true HPG or GV in any of the layers: the rules just tell us how they change from one layer to another. So we find it useful to choose an arbitrary HPG and GV in the lowest layer, and we label them HPGref and GVref to indicate that these are initially taken arbitrarily in the bottom "reference" layer. Using Rules 1 and 2 we can now construct the HPG and GV in each overlying layer. Using the data we can draw contours of T and PT in each layer. By Rule 3 the GV in each layer should be parallel to the PT contours. If this is not the case, we have the option of modifying our choice of HPGref and GVref.

Now that we have the rules, we can try to see if we can choose HPG(ref) so that we can align the full GVs with the isolines of PT in each layer. It is easy enough to satisfy this condition in the reference layer—all we have to do is chose the direction of the HPG(ref) to cut perpendicularly across the lines of constant PT; then the GV(ref) flows along them. We don't need to choose any particular magnitude of the HPG(ref), unless we are afraid of GV(ref)s that are too large. However, when we try the next layer up or down from the reference layer, we have to adjust the magnitude (but not the direction) of the HPG(ref) in order to ensure that at this new level the new GV is parallel to the lines of PT there. They are generally going to be in a different direction, if Figure 6.1 is any guide. Now, we have fixed both the direction and the magnitude of the HPG(ref). We cannot make any further adjustments because we have used up both the adjustable features of the system. The direction and magnitude of the HPG(ref) are all that we have to work with. This means that all the degrees of freedom of the system are used up, and when we come to the next layer

45

above or below, and look to see if the GVs there line up with the lines of equal PT, we may be in for a bad surprise: they may not line up. There is an inconsistency. Do we junk the observations? Do we throw out the rules? Or are we about to learn something? This is why science is exciting: you wrack your brains for a testable idea, and then you get so involved in it that you begin to hope that it won't be knocked down—even if you are about to learn something else.

The Chief has something to say. "I'm beginning to get an idea of what you are up to. But I want to see how your study of this particular block of ocean water fits into the whole picture. I get the impression that we're talking about just one piece of the ocean's machinery. So let me make a proposal. Suppose you were writing a maintenance or repair manual for the ocean, thinking of it as a machine. You'd start with some kind of exploded view showing all the parts—in this case blocks of water in different regions of the ocean. These might correspond to fireboxes, boilers, pistons, cylinders, reduction gears, and so on, in a man-made machine. You'd then go into some detail about how each particular piece works, and then you'd show how to put them all together into a working engine. How many blocks would you need to get a Mark I ocean?"

That is a novel question to me. I decide that probably about five choices would give a fair idea of the different parts of the machine. I choose Blocks A through E in the North Atlantic, omitting any reference to other oceans. The blocks that I chose as typical are shown in Figure 6.2. Block A is in the area of our present survey, but I did not draw it as a triangle, just a ten-degree square.

"I take it that there is something different about each of these five blocks. Do you have density data for any other of these boxes besides block A that we are working on?" asks the Chief. Yes, of course, there are a lot of historical data, nicely stored in numerical tabulations convenient for computing machines; but they are uneven in quality. Often the spacing is poor, coming from widely separated times. It is not sufficient to reveal the detailed variations in slopes and thicknesses of the layers from one level to another—layers from which we hope to diagnose the dynamical processes at work in the blocks. The noise level caused by eddies makes it tricky to evaluate the average slopes and thicknesses. Besides, in the case of our particular survey, we hope to derive even more than straight line approximations of layer levels—we are hoping for their curvature, for other purposes.

The Chief continues with his guidelines. "Is there any particular reason to prefer the triangular pattern of your stations over a rectangular one, like the blocks you drew?" No, I reply, except that it is more efficient to steam around, and gets more stations near the center. "Then," says the Chief, "I propose that in outlining our maintenance manual, we draw the blocks as rectangular as on the map here; then we'll be able to keep in mind a little better the distinction among the north, east, and other sides. Tell you what: I'll take this map with me and try to write out these ideas as a starter on our manual."

The Chief's bold approach surprises me. Like many scientists who fend off curious lay-people because they think the technical aspects are beyond lay comprehension, I had not expected the Chief to grab hold. He is forcing me to understand my own subject better. Next morning, after my morning watch, I discover that the Chief has left

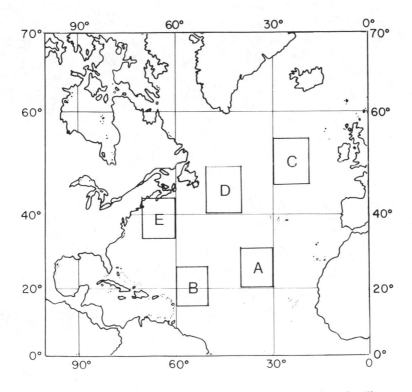

FIGURE 6.2. Five 10-degree squares in the North Atlantic are selected to illustrate different types of internal density structure. From these structures we can infer features of the hyodrodynamical processes occurring inside the ocean. We can also appreciate that the apparently undifferentiated water mass of the ocean can be envisaged as being made up of different parts, each of which operates differently and serves its own function in the overall machinery.

the following outline on my desk, entitled "Ocean Maintenance Manual":

Several different portions of the ocean's area are shown in Figure 6.2. Each is 10 degrees of latitude and 10 degrees of longitude. Because of the convergence of the meridians, the blocks farther north are narrower.

The density increases downward in each of these blocks. Although the real density increases smoothly, when thinking about it it's convenient to imagine that it increases downward by discrete steps (Figure 6.1). Each layer is then represented as being homogeneous in density and separated from its neighbors above and below by a sharp interface.

If one of the blocks were peeled off the globe, it would be thinner than a postage stamp, so to see the structure we need to exaggerate the vertical scale. In Figure 6.3 we show the upper kilometer of such a block with the vertical scale exaggerated about five hundred times.

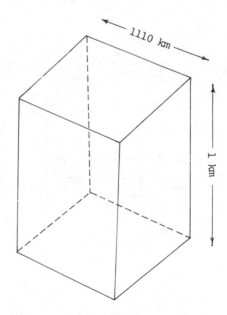

FIGURE 6.3. Perspective view of a block as seen by an observer high over the North Atlantic looking toward the northeast. The vertical scale is exaggerated five hundred times. Only the top kilometer is shown.

We can use oceanographic atlases (available in the ship's library), to draw the depth and slope of various density interfaces in these blocks. We can also show in plan view the contours of equal depth of each density interface. We show them for each of the five blocks in Figures 6.4 through 6.8.

Block A is located in the eastern part of the great clockwise subtropical gyre, where the surface currents flow generally southward. It is supposed to be in the region where the ocean is most strongly driven by the winds. (It is also the region we are studying on this cruise; I have drawn it square instead of triangular to make the displays somewhat clearer than is possible with a triangular prism as in Figure 6.1). Block B is in the westward-flowing portion of the subtropical gyre. The Gulf Stream cuts across Block E, so it is necessary to show some of the detail of this unusually narrow and swift current. Block D is a region of eastward-flowing current, which is also part of the subtropical gyre. Block C is part of another gyre to the north—the subpolar gyre that travels around the northern portion of the Atlantic in a counterclockwise direction. In Block C the surface current flows north and northeastward. The viewpoint of the observer is the same in these figures as that in Figure 6.3, from a point high above the sea and looking toward the northeast.

We are going to develop a picture of the currents in each of these blocks by using our three rules [presented earlier in this chapter]. From the slopes of the density interfaces (and the density contrast across them, which for the sake of simplicity we assume to be the same in all cases) we can obtain the change in HPG. This enables us to construct the change in GV. If, to begin with, we assume that the HPG and GV are zero in the deepest layer, then we immediately have the HPG and GV in all the layers above.

Now we construct the lines of constant PT in each of the layers. We then check to see if the GVs in each layer are parallel to these lines of constant PT. If they are not, then we are in trouble. Our first escape route is to accept the fact that our assumption that the HPG and GV are zero in the deepest layer may be wrong. We try various strengths and directions for the HPG and GV in the deepest layer; each time, we recalculate the HPG and GV of each of the upper layers, always trying to get the GVs (which are being varied with each recalculation) to line up with the lines of equal PT in all the layers simultaneously. If they do, then we have an indication that our three rules are working.

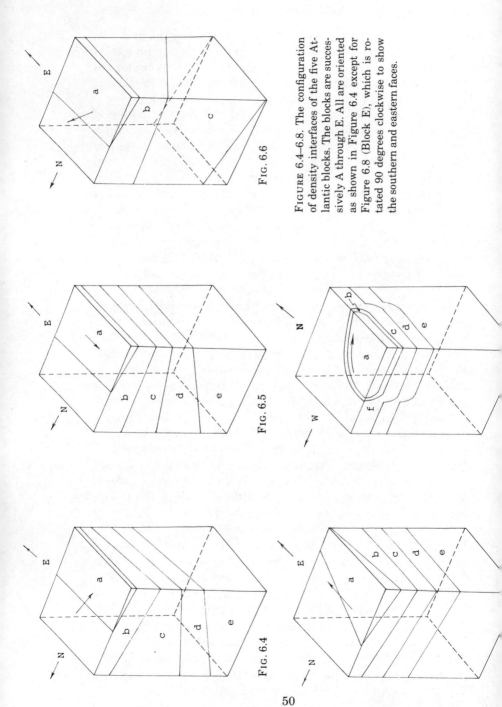

FIG. 6.6

FIG. 6.5

FIG. 6.4

FIGURE 6.4–6.8. The configuration of density interfaces of the five Atlantic blocks. The blocks are successively A through E. All are oriented as shown in Figure 6.4 except for Figure 6.8 (Block E), which is rotated 90 degrees clockwise to show the southern and eastern faces.

As I finish reading his text, the Chief comes in through the door, smiling. "That's as far as I could get—I think you'd better tell me about what comes next."

I turn to his picture of Block B, which is probably the simplest. You see, in this block all the interfaces slope either up or down toward the north. There is no slope at all in the east-west direction. Old oceanographers used to call this a case of *parallel solenoids*, and they thought all ocean currents had their slopes in only one direction. It was wishful thinking, but here in Block B it is very nearly true.

Therefore, if we start with zero HPG in the lowest layer, and go up across the density interface on its top, we will pick up an HPG pointing southward. As we ascend, the HPG will increase as we cross each interface, but it will always be southward. Even in crossing the top interface, which slopes toward the north instead of toward the south as the others do, the HPG will decrease a little but still point southward. So, when we calculate the relative GVs, all will go westward. The GVs in the top layers will be faster than those below. This is the region of the westward, so-called North Equatorial Current. It flows toward the west until it enters the Caribbean and Gulf of Mexico, and then goes northward in the Gulf Stream (which I will discuss more later). From the geometry you can see that the lines of equal thickness are also aligned east-west. Each is at a constant latitude, so the PT lines are also east-west, in every layer. Clearly, then, the GVs are flowing in the same direction as the PT lines, and Rule 3 is obeyed. We can assume that the HPG(ref) in the lower layer is southward (or northward, for that matter), and we get a westward or eastward GV(ref) in the lowest layer. When added to the relative GVs in the upper layers to get the actual GVs, the flow is still along the PT lines. So we do not have a unique solution to the GV(ref). We say such systems are *underdetermined*—they have many possible solutions that obey all the rules. On the other hand, if we are so bold as to assume that the HPG(ref) has an eastward or westward component and hence the GV(ref) has a south or northward component, then when added to the GVs of the upper layers they would all have this same meridionally directed component and would flow across PT lines, and Rule 3 would be violated. So, in terms of our three rules, we can say only that an arbitrary east-west velocity can be subtracted or added to the flow in all the layers. Oceanographers have tormented themselves for a century over their inability to find the full GV field in a block like Block B. They have called it the problem of finding the depth of no motion. Usually they have settled for choosing the depth of no motion deep

51

enough to make it plausible that there is only a small unknown current there. Logically sloppy, but unavoidable.

The situation is quite different in Block A (see the survey in Figure 6.1 or the Chief's sketch in Figure 6.4). Not all the density interfaces slope in the same direction. They all have a component of slope going toward the east, but some slope up toward the north, and deeper ones slope up toward the south. If we choose a zero HPG(ref) in the deepest layer, then we will also get a GV(ref) of zero there. Then, when we pass upward through the deep interface, we will get a relative HPG toward the southeast. We pass farther upward and add the HPGs; we find that the resulting HPGs gradually rotate around through the east until they point northeast. Starting at the top and going down through the various layers, the relative HPGs rotate clockwise and diminish in amplitude as we go down in the ocean through the various density interfaces. Since the relative GVs are always pointing to the left of the HPGs, the velocity has a spiral too. It starts at the surface, pointing southeast, and by the time the deep layers are reached it has diminished greatly and goes to the west. This clockwise rotation of the GV with depth (counterclockwise in the southern hemisphere) is called the *beta-spiral*. The main purpose of our surveys has been to prove its existence beyond a doubt, because there would always be some reason to question its existence as long as we had to depend on the sparse historical data in the area. I do not believe anyone will contradict me when I say that we have demonstrated the existence of this spiral in the HPGs and GVs.

On the other hand, we now must see whether the GVs flow along lines of constant PT in each layer, and, if they don't, whether they can be made to do so by some choice of the HPG(ref) and GV(ref) in the lower layer other than zero. It may be that no choice will avoid violating Rule 3 in some layer. In that case, we would have to find out what that tells us. Or it might turn out that there is a unique solution, and then we would have solved the problem of finding the level of no motion.

In any case, the twisting of direction of the HPGs from one layer to another presents us with a kind of Chinese puzzle with the layers. We want them to slip over each other, without breaking apart at the interfaces, and obey all three rules—or know the reason they don't. Something somewhere has to give. But I sense that we ought to postpone a further discussion of the blocks and the rules for another time. Let's go see what movie is being shown in the mess.

Seven / The Chief Tries His Hand

O UR SHIP is now about halfway down the third side of the triangle. We are rolling a little in the swell of the Trades, which is coming at us from the port beam. I check the chart once more because there is a seamount hereabouts—the Great Meteor Bank. I have already made this check on earlier cruises and know that there is nothing to worry about, since our course is to the west of it. But being a worry-wart, I just have to look once again.

We have seen other shipping during the cruise. One afternoon the Mate saw a schooner lowering her sails—a signal of distress. We could not raise the crew on the UHF radio, so we cruised over to the ship to get within hailing distance. It was a yachting party from Britain, heading to Antigua. They had run out of cigarettes and wanted to borrow some. Our skipper is a good sort; he passed over a dozen cartons in a plastic bag by heaving line and forebore to scold them for misuse of the distress signal. They tried to reciprocate by passing over some beer in a cardboard case, which dissolved in the water, and in their excitement almost ran their yacht into our hull. That would have been serious indeed. Thank goodness for storms at sea that keep the ocean fairly free of tourists such as these.

THE CHIEF comes around with a sheaf of papers containing his attempts at some of the calculations we talked about last time. "You know," he says, "the density structures in those blocks I drew up the other day are so complicated that I decided I better start with some really simple ones, ones with only one interface or none at all—just homogeneous water." I comment that that is the true scientific approach: always be ready to simplify; never carry more baggage than you have to; travel light. I am curious to know what he has decided to use as a simple model.

He pulls up a chair and offers me the following explanation: "The simplest case I can think of is an ocean with a flat bottom and homogeneous density. In that case the HPG and GV have to be the same at all depths. There are no density interfaces, so we are free to choose any magnitude and direction of the HPG, and then, involving Rule 2, we have a corresponding GV. This HPG is just like the HPG(ref) in the more complicated problems. Up through Rule 2 there isn't any restriction on the HPG and accompanying GV. But when we come to Rule 3 we notice that the PT lines are east-west, so we can't have any north-south GV at all. We're restricted to eastward or westward geostrophic flows in this kind of ocean.

"This isn't so very strange: How can columns of water grow or shrink in vertical height as they must do if they move north-south, when the water everywhere has the same depth? I ought to be careful here to distinguish the large changes in vertical height that are required to keep the potential thickness, PT, conserved, when compared to the small variations in surface height that would be needed for the various choices of HPG. One also needs to remember that even these small height differences are not available for PT shrinking because the GV is normal to the HPG.

"The easiest way to free up the block so that water can flow through it at any speed in any direction without breaking the three rules would be to give the bottom a slope down toward the north so that the total depth increases proportionally to the sine of the latitude; then the potential thickness would be uniform throughout the block. That would permit water to flow in any compass direction without having to go from one potential thickness to another and without breaking Rule 3. That seems to me somewhat like cheating."

The Chief pauses, but he has more to say: "Then I asked myself, how could I get some north-south GV in this setup without resorting to a sloped bottom? It occurred to me that if I were allowed to add water to the top of the ocean—maybe by rain, for example—then the main column of water could shrink as it moves south, but the total height could remain the same because of the addition of water on the top. I like to call that extra amount of rainwater added to the top of the column a *rain hat*. The farther you want the GV to carry the column toward the equator, the shorter the unhatted portion of the column must become, by Rule 3; but my Rule 2 says that the total column height must remain the same, so the rain hat has to get higher to make up for the shrinking part of the original column of water. The north-south speed would have to be proportional to the intensity of the rain, but the east-west component of the velocity could be anything at all. Do you think that would work?"

Yes, I tell the Chief, that is the essence of the matter, although, when you consider the kinds of ways that water might be added to the top of a moving column, rain probably is not very important. The excess of rain over evaporation over the course of a year seldom exceeds one meter a year, but the winds in the trade-wind region manage to push down a wind hat at a rate of about 30 meters of surface water a year. That turns out to be enough to allow columns in the upper 800 meters of the subtropical Atlantic to move southward with GVs that can carry about 30 megatons per second of water southward. The same volume moves northward through the Florida Straits and is known as the Florida Current, which is the source of the Gulf Stream. We probably ought to delay going into the rain-hat business for awhile and simply say that the strict application of the three rules to the simple case you invented restricts the system to east-west flows. To get north-south flows in this homogeneous flat-bottomed ocean, you have to find a way around Rule 3. This means that Rule 3 is a pretty restrictive condition.

Then the Chief pulled out a drawing. "That's what I thought, too. So instead of trying to consider the full range of depth, I decided to consider a few sloping submerged layers—much as you are doing here in the ocean. My aim was to try to find some configuration of layers that would permit free north-south flow obeying the three rules, even if only a limited range of depth. In this sketch (Figure 7.1) I've drawn two layers a and b bounded by three interfaces 1, 2, and 3, all sloping

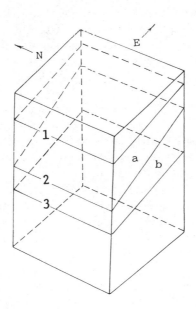

FIGURE 7.1. The simple block that the Chief used for his discussion. There are two layers of density a and b, bounded by interfaces that slope down toward the west.

downward, toward the west. It is evident that layer a has a southward GV_{ab} relative to whatever the velocity is in layer b. I first drew the slopes of all the interfaces the same, so the layers a and b had uniform thickness. But then I discovered that this would imply that the PT (potential thickness) contours ran east-west in both layers. This would mean that neither layer has a north-south component of velocity, and hence would contradict the required southward GV_{ab}. That is why the middle interface is drawn with the greatest slope and the layer thicknesses are not uniform. Contours of equal thickness run north and south; contours of equal potential thickness are twisted away from north in opposite directions. Now there is a hope that we can avoid a contradiction, and that if there is a unique solution we will actually have found that long-lost reference level, or "depth of no motion." Maybe we will even have a rudimentary spiral in the velocity.

"Then I drew up some plan views that show the lines of constant depth D of the interfaces (Figure 7.2). They are straight north and south, as shown in the panel on the left. Then for each of the layers a and b I sketched the lines of constant layer thickness T. They are also straight north and south. The upper layer gets thicker westward, and the lower one gets thicker eastward. The directions in which they get thicker are indicated by the arrows in the $+$-direction.

"Then I drew two more panels for these layers to the right to show the lines of constant potential thickness (PT). These are tilted because of the latitude effect in Rule 3. As you go southward along one of these lines, the layer actually has to be thinner by the sine of the latitude. For example, in layer a you have to move a little east of south to keep at the same PT.

"Looking at the left-hand panel, we see that the slope of the interface requires, by Rule 1, that the HPG in layer a relative to that in layer b must be eastward. I indicated this quantity by the symbol HPGab, which is supposed to be read 'HPG in a relative to that in b.' Then, by Rule 2, I found the GV in a relative to b. It is the arrow labeled GVab in the bottom left panel. Now, Rule 3 demands that the actual GV in layer a (let's call it GVa) must be parallel to the PT lines in a, and that GVb must lie parallel to the PT lines in b. Alongside the arrow GVab I drew these two required directions from the plots of PT above. Now I come to the climax: in the last panel I simply resolved GVab into two components, GVa and GVb, that lie along the required directions. Or to put matters another way, I find the right length of arrow GVb along the PT lines of layer b, which, when added to the known GVab, also gives us an arrow (GVa) that lies along the PT lines in layer a. I think that these are the two GVs in this block that satisfy all the rules, and that they are the only ones.

"So my hunch that putting density interfaces into the block would allow us to get some north-south flow without breaking Rule 3 seems to be right. But from one point of view it hasn't exactly freed things up. Now there is only one solution, whereas in the homogeneous ocean block there were any number of purely east-west solutions."

Good heavens, I think, this is some Chief! I hope he doesn't quiz *me* about diesels. Anxious to keep ahead of him, I tell him that when more than two layers are put into the blocks, things get worse, and that it would be harder to keep Rule 3; but in exploring what modifications to make, we are going to learn something about how the ocean machine is driven.

For almost a century, oceanographers have puzzled over how much the ocean is driven by the stress of the winds and how much by horizontal variations in density caused by the sun's uneven heating and

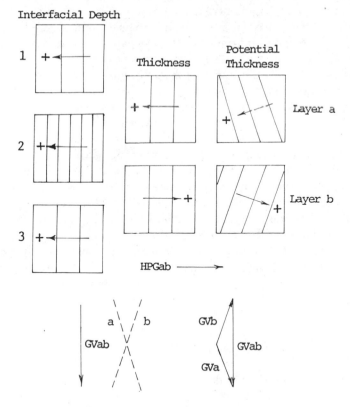

FIGURE 7.2. The Chief's diagnosis of the application of the three rules to the imaginary block he invented in Figure 7.1.

by the cooling of sea water in high latitudes. Evaporation and rainfall can affect the salinity of the ocean water, and this in turn affects the density.

In the 1870s the English journal *Nature* carried a series of letters debating these two causes of ocean circulation. Dr. Carpenter, who was considered perhaps a little pompous (the young men on HMS *Challenger* had a parrot whom they taught to say, "Aha, Dr. Carpenter, F.R.S.") held to the view that density differences drive ocean currents. His adversary, Dr. Croll, was at the time a janitor at the Anderson Technical College in Aberdeen, and he propounded the view that currents are driven by the wind. The correspondence in *Nature* grew heated and personal until finally a third letter-writer called for a suspension of the interchange; but he withheld his name so that "it wouldn't influence the matter." Alas, we will never know who this exalted personage was—the files of *Nature* were destroyed in the Blitz. For many years now the belief has grown that wind-stress is predominant in driving ocean currents. This is perhaps an example of the stultifying influence of textbook writers who depend on published scientific papers for all information. If these papers tend to be one-sided, then of course the text is, too. This gives students a wrong impression—they see only those parts of the problems that have been worked on. When someone asks me why my own early models were all wind driven, if I was not confident that wind is the main driver of ocean circulation, I can only answer that it was because the wind-driven models were easier to formulate. My response reminds me of a Japanese friend who, though a Buddhist, was married in a Shinto shrine. I asked him why. He replied, "It was cheaper." Similarly, our early preoccupation with wind-driven models can perhaps be justified because they are easiest. But we do need to do more thinking about density-driven models. Diagnostic studies like the one made by the Chief help stimulate our thinking in this direction.

It seems to me that the best way to show the Chief that adding more layers to blocks does not add more degrees of freedom is to go over still another example with him. So I draw a box with a flat bottom (Figure 7.3). The two interfaces are numbered 1 and 2. There are now three layers *a, b,* and *c*. This is much like the previous block and is a simplified form of the earlier Block A (Figure 6.4). In Figure 7.4 I go through the same arguments the Chief used. First, on the left upper panels labeled 1 and 2, I indicate the depth of the two interfaces, with the arrow pointing to the deeper side. Then, in the next panels, the three vertical ones under T, I indicate the contours of equal thickness; none have the same direction of increasing thickness. Then, in the three panels directly under PT, I indicate the contours of equal potential thickness. They differ from the T-lines because of the sine-of-latitude division. Once I have finished these, I go down to the bottom line of panels. Because of the direction of the slope on interface 2, the HPG in layer b relative to that in layer c is toward the east. I label it as HPGbc. This means that the GV in b relative to that in c is southward, shown as GVbc in the next panel to the right. This panel also contains dashed lines indicating the alignment of the PTs in layers b and c, and they are accordingly labeled.

One must now make a vector construction that will start with a GV in layer c, parallel to the T-lines in c. It can be called GVc and, when added to the GVbc, results in a GVb that lies along the PT lines in layer b. This construction (see the bottom right panel of Figure 7.4) shows that it is possible to tie layers b and c together using all three rules and that there is a unique result (shades of the unknown level of no motion). Now, using this value of GVb, we can add GVab to it, and we will see that it is not parallel to the PT lines in layer a. Therefore, we cannot construct the three-layer model so that it is consistent with the three rules, unless there are very special circumstances. The slopes and thicknesses have to be set up just right. This is the wonder

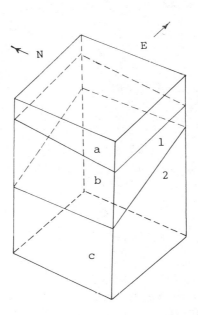

FIGURE 7.3. A more complicated case of Figure 7.1., with three layers.

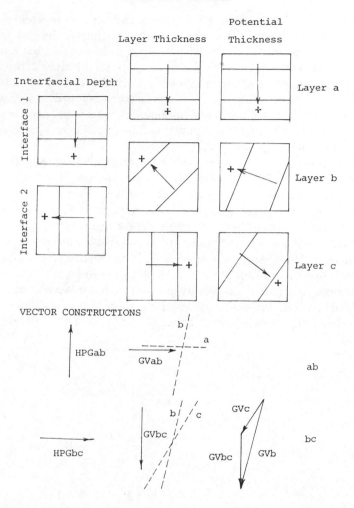

FIGURE 7.4. Diagnosis of the more complicated case in Figure 7.3, showing how a density structure can violate the three rules.

of the beta-spiral. It is the ocean's attempt to obey the rules with as many layers as you please. In principle, the inverse problem of determining the absolute velocity field from regions of depth where the three rules may be expected to apply is vastly overdetermined. When we attempt to perform the inversion on real data, we find that it is barely determined for reasons that we will have to explore further later—maybe when the ship gets into Cadiz and we have some free moments to contemplate things.

It is not quite fair to use this particular example by the way. No one really expects the top layer to conserve potential thickness. It has water added to the top in the form of wind hats—forced downward from a shallow, wind-driven layer. So in all likelihood, if we took this schematic model seriously, we would accept the southwestward GVa and calculate how quickly the wind hat has to grow to offset the violation of Rule 3 in layer a; then we would tell the meteorologists that we had found a check on their calculations of the wind-driven downwelling from the top mixed layer.

The Chief looks at me quizzically. "You seem to be using the word 'determined' in some kind of technical sense. You've said 'underdetermined' and 'overdetermined' several times now. Do you want to play fair and explain what you mean to me?" I can see that some jargon has crept in. What I mean, Chief, by the word *underdetermined* is that there is not enough information from the data to figure out the answer. *Overdetermined* means that there is more than enough, and opens the possibility that you get different answers when you choose different data. For example, if you are on the bridge ("God forbid," growls the Chief) and taking bearings on landmarks and lighthouses, you need two independent bearings to get a fix for your position. If you have only one bearing, you have a line on the chart that you know you are on, but you need another line to get an intersection and a fix. The coastal navigation problem is determined if you have two bearings, underdetermined if you only have one.

"O.K.," says the Chief, "and when you have more than two bearings, the problem is overdetermined. But if you've made some mistakes, you'll get different answers depending on which bearings you choose to use—not unlikely on the bridge, in my opinion—so either you average them out, or you look for some more subtle mystery in the whole business, like horizontal refraction, or something." That seems like a good enough approximation for my money, and I leave it at that.

It appears to me that the Chief has done enough for the day, so I make unmistakable signals that we had better continue another time. The Chief scratches his head behind an ear and asks, "How did you come up with all these funny ideas?" So I tell him a little about my sabbatical five years ago, in 1976, when I visited Kiel, Germany, for a few months, and thereby got away from teaching, project management, and the multitude of distractions that help ruin a man's scientific productivity. It was mostly dark, cold, and wet. The Institut für Meereskunde is right on the harbor, and I walked to work in the morning and home in the evening along a promenade from which the whole busy harbor was visible—shipyards burning with welding torches, huge liquid gas tankers being built just across the water, and busy little ferries carrying passengers. In contrast to my office at M.I.T., or for that matter the one at Woods Hole in the new Clark building, far removed from the sea, in Kiel marine activity was immediately at hand. One simply could not help thinking about the ocean, ships, instruments, and oceanographic problems. The air I breathed was alive with them.

My German hosts saddled me with no duties, so all I had to do was think. I thought about things that interested me, particularly the problem of getting absolute geostrophic velocities. A vague picture began to form in my mind, and when I was ready to try some calculations I met a bright young German oceanographer named Fritz Schott, who understood right away what I was trying to do. His superior computer and statistical skills made all the difference. His collaboration gave rise to a sense of shared excitement, and we were soon phoning each other up at all hours with new ideas. That was when we wrote together our paper on the beta-spiral, using historical data.

Two years later, in 1978, a meeting was held at the University of Rhode Island to plan new, large cooperative programs. I toyed with the idea of trying to launch a large beta-spiral sea survey that could have been called the "Sverdrup Experiment." It could have led to one of those excruciatingly boring think-tank experiments with dozens of meetings, an executive officer, and all of that. If that happened, I suddenly realized, I would be right back in the administrative role that I was determined to escape once and for all.

Fortunately, I came to my senses and decided to reserve the idea for a modest personal investigation. So I started out small, more or less on my own. I was joined by Dave Behringer, who had been working independently along similar lines, without Schott's and my knowledge. On one cruise Larry Armi also joined me. Between us, Dave Behringer and I have already conducted four separate surveys in the triangular area of the beta-spiral. Which brings us now to July 1981. Here I am in my cabin on the research vessel *Atlantis II* during the fifth and final cruise for this project. The nearest land is some 800 miles to the southeast—the Cape Verde Islands. We had completed the outer perimeter of the triangle and are now filling in the center with some extra stations, for which we fortunately have spare time.

We are coming to the end of our survey, with only a few more stations to go before heading east to Cadiz. When I enter the main laboratory that evening I find the Chief musing over some of our plots and sketches. I think his favorite is still the old one that we had completed over a week ago and posted on the side of a refrigerator: the one showing the density surfaces plotted up in a very heavily smoothed fashion (Figure 6.1). He turns to me and asks: "What have you finally decided about this one? It's a little too complicated for me to work out like the other examples."

Well, I reply, it's not very different from what we found on the other four surveys—and that in itself is reassuring. It means that there is a permanent, stable structure here—may be not as stable as the Rock of Gibraltar, but not wildly varying in time either. When we tried out the diagnostic calculation to see if we could line up the GVs with the PT lines in all layers, we had a fair success. The layers that seemed to govern the spiral most strictly were layers b and d. We have to discount layer a because it probably is fed from the top with water pumped downward by the wind, so Rule 3 does not work there. Then layer e has to be left out because it is actually very deep, extending down to the ocean bottom, where the irregular bottom topography and friction may upset Rule 3. It would have been nice if we could have used layer c as a check on the reference level determined from layers b and d, but unfortunately the PT is so uniform in layer c that we cannot get a good map of the PT lines there to compare with the GVs in that layer. The relative velocities computed are therefore not really checked for consistency with a third layer, but they are consistent with layers b and d and lead to reasonably small velocities in the deep layer e. Using them, we can also compute the downward velocity needed from the wind-mixed layer at the top of layer a. This velocity turns out to be between 30 and 40 meters a year and agrees with that computed by meteorologists independently from measurements of the average wind patterns at the surface.

But we have been able to do more than make this diagnostic calculation with our data. The smoothing of the data in Figure 6.1 was so severe that all that remains are the mean depths and slopes of the surfaces. However, there are enough data in our surveys to permit somewhat weaker smoothing, so that we can get the curvature of the property surfaces, which has permitted us to make estimates of large-scale mixing by turbulence due to eddies. I would not have had the courage or resolution to tackle this task without the very happy collaboration with Dr. Laurence Armi. He is an engineer as well as a scientist, and he has a firm belief in extracting as much as possible out of data. Incidentally, the velocities in the spiral are quite small. Figure 7.5 shows the velocity arrows (GVs) at depths of 0, 200, 400, 600, 800, and 1,000 meters on one of the surveys. You can see that none of them are

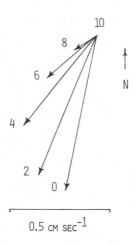

FIGURE 7.5. The observed rotation of the GV with depth (in hundreds of meters depth) as obtained by the diagnosis of one of our beta-triangle cruises.

greater than one centimeter a second (half a mile a day). At that velocity it would take a particle ten years to move 2,000 miles. That gives a fair idea of how slowly the water revolves at depth around the main oceanic gyres. When compared to the angular velocity of the earth, one rotation per day, the spin of the gyres relative to the earth is only 1/10,000th as much. In other words, the pressure anomalies associated with ocean circulation deviate only very slightly from those of the overall equilibrium of the rotating earth system.

IF I WERE asked, "What do you think was the most important thing that you determined from your five surveys of the triangle region?" I would answer that we were the first to establish that there really is a clockwise rotation of the density slopes with depth in a northern hemisphere subtropical oceanic gyre. We were able to get an estimate of how steady this configuration is in time. Before our study oceanographers had, it is true, noted that the north-south slope tends to reverse with depth, but they had not noticed the important rotation that is so crucial to locking together the Chinese puzzle, so that for a given configuration of density, the water can slip through it only at one latitude. Our beta-cruise data have enabled us to do the diagnostic studies of real data and to demonstrate that the three rules actually do, to a measurable degree, apply in mid-ocean.

These results are most dramatically exhibited when the diagnostic calculation is carried out for a large number of density interfaces simultaneously, and when the solution is found for the least sum of the square of the residuals (badness of fit). As our second most important accomplishment, Schott and I discovered that the smallness of this minimum of the residuals is very sensitive to the latitude used in the calculation. We discovered this sensitive tuning to latitude accidentally, because at one stage of our studies we were unintentionally using the wrong value of the latitude for the middle of the triangle (actually we had the wrong sign and computed for the southern hemisphere first). Figure 7.6 shows how sharp the minimum of the residuals is at the correct latitude. It is taken from our first paper, where we used historical data only.

I should also like to be able to claim a third accomplishment: we de-

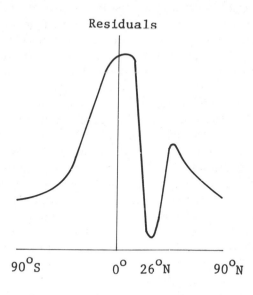

FIGURE 7.6. Least-square residuals in the beta-triangle area calculated from the observed density structure but using various wrong central latitudes (as well as the correct 26°N). The very sharp minimum residual that occurs when the latitude is chosen correctly indicates that the internal density structure of the ocean is finely turned by the dynamics of Rules 1-3.

vised a practical method for determining the absolute velocity field—the solution of the perennial problem of determining the "depth of no motion." But this we have done only in principle, not well enough to be of practical use to people interested in the direction and speed of deep and abyssal flows in the ocean. Evidently the unsteadiness of the gyre, the noise due to eddies (with a radius near 40 kilometers), and the probable fact that the rules are not exactly obeyed prevent us from reaching this goal.

On some of our beta-spiral cruises we gathered water samples for Dr. William Jenkins, who had developed a clever way to analyze them for tritium and helium-3. The tritium in the upper layers of the ocean was deposited during the hydrogen bomb tests of the sixties; the helium-3 is a decay product of the tritium, so it labels the water layers with little clocks. Jenkins got velocities of water flow quite similar to our dynamically determined ones.

I HAVE already reached my sixtieth birthday, and, except for the Chief, I am the oldest man aboard. Between stations, my cabin is a good place for contemplation. At this time of year, the winds and the seas are both moderate. It is not oppressively hot. If one can ignore the deafening chatter of the pneumatic chipping hammers by day, and the loud whine of the blower with defective bearings close to one's head at night, and brace oneself against the mild but irregular, wallowing motion that is beam-end to the Trades, the cabin is an ideal workplace. There are no visitors, no telephones to ring, no temptation to drive down to the drugstore for an idle purchase. In fact, there is very little else to do but think and work. And the urge to express myself, to communicate my own view of the ocean—as limited and partial as it must be—is irrepressible. Of all the wonders of the universe, surely enjoyment of exercising our minds must be the greatest. Our own personal lumps of grey matter—the ultimate in personal computers—enable us to play computer games with the physical universe. Could any child ask for a better toy?

Station work and some instrumental breakdowns keep me awake most of the night. I don't retreat to my berth until about six in the morning, but I cannot sleep. Thoughts of my discussion with the Chief are running through my head, and I keep wondering whether somehow I had presented something wrongly. A scientist is on pretty safe ground as long as he stays close to the mathematical formalism. But when he tries to discuss the physics in simpler, more intuitive terms, using only simple geometrical constructions, there is a danger that he will commit some error of explanation. Had I overlooked something in the way we were applying the three rules? Did the box-type model conceal some subtle contradiction of which I was not aware? (Several years later, while working with Nelson Hogg on some boxlike models, we did indeed discover a subtle contradiction in one of the models that was in this chapter. It has been corrected in the text.) You remember the old prayers for those in peril at sea? Perhaps there should be a special prayer for those in peril of trying to explain the sea.

Eight / Exceptions to Our Three Rules

W HEN WE discover, through our diagnostic studies of the density structure in various blocks of the ocean, that the three rules in Chapter 6 do not rigorously apply in all layers, we have to examine the rules. On physical grounds, our suspicions center on Rule 3. Table 1 lists the rules again, plus three likely codicils to Rule 3. The first codicil is an exception that allows us to include the effect of the shallow drift-currents produced by the wind: we call this modified version Rule 3W ("W" stands for wind). The second allows us to include the effect of changes of density of the water by heating or cooling or by changes in salt concentration by evaporation or precipitation: we call this codicil Rule 3B ("B" stands for buoyancy). The third modification of Rule 3 allows us to include the possibility that our columns of water within the current take on a large spin about a vertical axis relative to the earth, and approach even the spin of the earth itself. This spin is called *relative vorticity* and so this codicil to the rules is called Rule 3RV ("RV" stands for relative vorticity).

In Figure 8.1 we depict a wind stress acting toward the east at the top of a block of ocean. In a steady state this force has to be balanced by something. The noted Arctic explorer Fridtjof Nansen noted that ice in the top of the ocean seems to drift to the right of the direction toward which the wind blows, and he suspected that the water in the top layers of the ocean does, too. He was puzzled and thought it might be due to a Coriolis force (CF). He communicated this idea to the young V. W. Ekman, an oceanographer, in 1902, and within a short time Ekman had worked out his theory of wind drift. It turns out that the viscous shearing stresses in the upper layers of the ocean turn sharply to the right with depth, and the total stress is confined to a shallow layer and does not penetrate all the way to the bottom as it might in a nonrotating reference frame. The confined stress layer communicates the wind stress to water in the upper hundred or so meters, but not below, so by itself it cannot balance the wind stress. The actual balance is due to a Coriolis force, associated with a mean current drift to the right (in the northern hemisphere) of the wind stress. The CF is to the right of the transport (which we can call ET, for Ekman transport). Two right angles add up to having the CF opposed to

Table 1
The Three Rules and Their Three Codicils

RULE 1. The horizontal pressure gradient, HPG, increases as we ascend through a density interface from one layer of homogeneous density to the one above it by an amount proportional to the product of the density contrast and the interfacial slope, and the increase is in the direction of the upward slope.

RULE 2. Within a layer of uniform HPG, the geostrophic velocity, GV, is also horizontal and independent of depth and is directed to the right of the direction of the HPG (in the northern hemisphere). The amplitude of the GV is proportional to the amplitude of the HPG divided by the sine of the latitude. A corollary is that the change of GV as one ascends across a density interface is to the right of the change in HPG and proportional to the amplitude of the change in HPG divided by the sine of latitude.

RULE 3. Because the GV is independent of depth within an individual density layer, the water within a layer moves as a vertical column, without tilting with respect to the local vertical. To conserve mass, any movement in the north-south direction requires that the vertical height of the column vary as the sine of the latitude. The height of the column is simply the thickness of the layer, T. Therefore the GV in the layer must be parallel to contours of a quantity we call potential thickness, PT (where we define PT as T divided by the sine of latitude). Because a column follows a contour of constant PT, its height (T) is proportional to the sine of the latitude.

RULE 3W. The change in height of a column following contours of constant PT can be compensated for by the addition of water vertically at the top surface from a convergent wind-driven layer. We call this addition Ekman pumping. The amount added is an *Ekman hat*.

RULE 3B. Similarly, buoyancy sources at interfaces can require fluxes of water across the interfaces, compensating for vertical height changes in the rest of a column, and adding water at its bottom. The amount added is called a *buoyancy boot*.

RULE 3RV. If the horizontal scale of a strong current is small, relative vorticity can be important and modifies the simple Rule 3. The height of a column then remains proportional to the sum of the Coriolis parameter and the relative vorticity. When the relative vorticity is small, as it tends to be for the large gyres in mid-ocean, it reduces to Rule 3.

the wind stress, so we have a balance of forces between the wind stress eastward at the surface, and a CF averaged over the shallow wind-driven layer within which the ET is occurring.

Now, wind stress is not uniform over the ocean. Figure 8.2 shows three adjoining blocks of water in the middle of the ocean. The wind is blowing toward the east over the surface of the northernmost block;

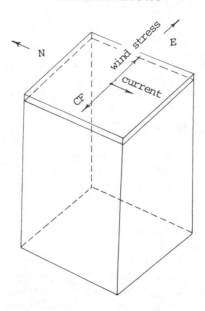

FIGURE 8.1. A block with a very thin surface layer that is acted upon by wind stress (in this example toward the east). This stress is not communicated to the water below because of the low viscosity of water and is entirely balanced by a CF acting toward the west, which comes from a thin drift current heading toward the south—that is, to the right of the applied wind stress. This thin current is the so-called Ekman layer, and the amount of water flowing is independent of the thickness of the layer or the magnitude of the viscosity or enhanced effective viscosity due to surface turbulence. This remarkable phenomenon was first understood by V. W. Ekman in 1902, and for many years was one of the few quantitative physical ideas in oceanographical knowledge.

no wind blows over the middle block; and the wind is toward the west over the southern block. This is a very crude representation of the distribution of the winds over the subtropical gyre: Westerlies in the north, and Trades in the south.

Both ETs are directed toward the central block, and if the water in the top wind-driven layer is not going to pile up, it can only escape downward. The little vertical arrow W_e denotes this downward flow; we call it an Ekman pumping. This water enters the friction-free layer of water just beneath, in which Rule 3 would otherwise apply, and adds some water at a constant rate (in our beta-spiral region, Figure 6.1 or Block A in Figure 6.4, this rate was found from our diagnostics to be about 30 meters a year). So each column of water in a layer just underneath the wind-driven Ekman layer may have its top added to (gain an Ekman hat) or cut off (Ekman guillotine) if the Ekman layer sucks water upward instead of pumping it downward. In this way we devise the idea of Rule 3W, which is important because it is the rule by which the wind enters into the theory of ocean currents as a driving mechanism.

Now we come to the next codicil, Rule 3B. This exception is confined to relatively shallow layers, too, often not much deeper than 500 meters. This occurs because of the uneven heating of the earth by the sun. Recent satellite measurements, using radiometers that measure the sun's heat coming into the earth and the earth's longer-wavelength infrared heat radiating away from the earth, confirm and quantify what we have known to be the case for a long time: there is a net radiative input of heat at low latitudes and a net radiative loss of heat at high latitudes. For the earth to have a steady climate, this excess of tropical heat has to find its way poleward via air and water currents. Since 1945, when the grid of upper-air observing stations was finally complete enough in the northern hemisphere to provide accurate statistics of wind directions, humidities, and temperatures at all

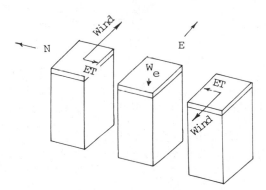

FIGURE 8.2. Three adjoining blocks of water with different applied wind stresses. The northern block has Westerlies blowing over it, as in Figure 8.1. The southern block has a wind stress toward the west (to model the trade winds). Its Ekman transport (ET) is toward the north. Both ETs are therefore directed toward the central Block 9, where the wind stress is assumed to be zero; this surface water, which would otherwise tend to accumulate, is forced downward into the top layer of the central block. The downward velocity is called Ekman pumping, denoted by w_e. Vertical Ekman pumping of water is the main way for winds to drive deeper ocean currents.

atmospheric altitudes, it has become possible to make reliable estimates of how much of this poleward heat transport is carried by the
air. It turns out that it is only about half that required for the radiation balance, and presumably the other half must be carried by ocean
currents. However, we have far too few and infrequent oceanic data on
currents and temperatures at all depths to make reliable, observa-
tionally based estimates of the ocean's heat transport, except in a very
few especially favorable locations.

If we try to include heat gain from and heat loss to the atmosphere
in our blocks of ocean, we will want to be able to convert water from
one density to another, and this means we will want to permit flow of
water across interfaces—at least between the uppermost two density
layers. For example, meteorologists tell us that in Block C, off Ireland,
there is a continuous loss of about 50 watts per square meter (this is
how Europe stays warm). If the temperature difference of the two uppermost density layers amounts to, say, 5°C, our block of ocean would
show it as a flux across the top interface from warm to cooler layer of
several hundred meters a year. This is a taller hat than the wind-
caused Ekman hats. The flow across the interface adds water to columns in one layer (buoyancy hats) while it cuts it off (buoyancy guil-
lotines) from the other layer. Rule 3B is the modification of Rule 3 that
allows for this process.

When we do the diagnostic construction for Block C (Figure 7.7), we
discover that the three-layer system is jammed. From meteorological
data it appears that in this region Ekman hats are not in style, but
buoyancy hats probably are. So we can get out of the jam by invoking
Rule 3B in the upper two layers instead of Rule 3W for the uppermost
layer alone. Then we get an estimate of the interfacial flux as well as
the climatologically important net heat loss from the ocean's surface
off the Emerald Isle.

The third codicil to Rule 3 comes about in special regions of the
ocean where the density slopes are locally very great, as in the Gulf
Stream (Block E, Figure 7.9). Associated with the steep slopes are
large HPGs and large GVs (the GVs in the Gulf Stream are 100–300
times greater than the slow GVs in Block A). Because the region of
high GV is narrow, there are regions of strong shear along the edges
of the Gulf Stream. Since the Gulf Stream's width is only one hun-
dredth of the width of the gyre, and the currents are one hundred
times stronger, the horizontal shears on its edges are ten thousand
times greater than those in the main gyres. The spin about a vertical
axis associated with these shears is therefore comparable to the spin
of the earth on its axis.

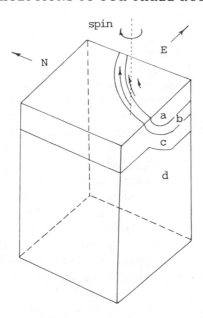

FIGURE 8.3. This block shows the Gulf Stream flowing through it. The currents are so large and narrow that there is a perceptible spin of the water around a floating vertical axis. This spin almost cancels out, locally, the rotation of the earth.

Figure 8.3 is a sketch of a four-density layer segment of the Gulf Stream where it flows northeast. Currents fall off so rapidly toward the southeast that if we were to set out a pattern of floats in this region, the pattern would rotate clockwise almost once a day. In other words, the spin of the water around a floating vertical axis in the southeastern flank of the Gulf Stream would almost be great enough to cancel out the rotation of the earth. Obviously, our whole line of thinking, which is dependent on the idea that pressure and velocity depart from the equilibrium of a uniformly rotating, self-gravitating spherical globe in absolute space, may be in peril. When local spin (we can also use the term "relative vorticity" for spin) is large, we need to modify Rule 3 to a new form, called Rule 3RV. Because relative vorticity is important only in special, small regions of the ocean, such as the narrow Gulf Stream, and apparently not important in the broad open reaches of the oceanic gyres, I will not pursue our discussion of the effects of RV here.

I AM A LITTLE worried that the Chief's patience is wearing thin, and

I ask him whether I am going into too much detail. "No," he says "I don't think so. In fact, I was rather hoping for these modifications of the rules. When you started with them, I thought to myself that they were altogether too rigid. Parts of machines, such as crank shafts and gear trains, the basic structural pieces, are tightly constrained. They have no 'give' in them. They don't really play much of a role in a machine besides pass power on from one place to another. The real things that do the work are the fire boxes, the combustion chambers at one end of the machine, and the screws (propellers) at the other end. I didn't see the input and output in the three rules, so it came as a kind relief when you found you needed them. What surprises me is that you were able to deduce it from the density structures themselves. I see Rule 3W as the fuel injector for wind work, and Rule 3B as a heater/condenser, and I guess Rule 3RV is some kind of safety valve or exhaust pipe, where the fluid finally finds a way to break loose of those really severe constraints of having to keep its spin in absolute space so close to that of the rotation of the earth."

Nine / Various Kinds of Blocks

W E H A V E finished the last station in our survey, and have steamed eastward around the light on the little island Las Puercas into the Bahia de Cadiz. This time we have a good berth—the best in town—right in front of an excellent quay-side cafe. The Captain—bless his soul—has made arrangements for us to have our meals paid for at the cafe. It is only thirty feet away from our gangplank, and even the watch is sitting there, drinking beer. I like a captain who is willing to take his chances with the accounting office at home.

The Chief and I are going to part company. I am going to fly to Seville, London, and then home. He is going to stay with the *Atlantis II* for a couple of more legs. The time has come to sum up our discussions. As the Chief says, "Now that you've told me about the parts of the ocean engine, let's see how they fit together." The coffee has been cleared away, and there seems to be an endless supply of Spanish brandy. The sun has already set over the Atlantic across the little peninsula on which Cadiz stands. The air is soft, and one thinks of that peaceful night four hundred years ago when the sea dragon Drake was approaching this same harbor to singe the beard of the Spanish King Phillip, to set the galleons in this port ablaze.

I brought the Chief's manuscript, "Maintenance and Repair Manual for the Ocean," along with me so that we can use some of the blocks he had drawn.

LICKING my pencil, I get to work trying to explain how oceanographers think about putting the blocks together. The Chief says, "They have to go together because we got them by cutting up the observed field of density in the ocean." I have to concede that what I meant was that if we imagine them to be cut out of the ocean like a jigsaw puzzle and then shuffled, there would be certain features that would help us presort them. For example those with low surface densities and high temperatures would likely belong in the tropics. The actual fit would only be clinched if the density surfaces matched at the common faces when finally assembled. "That isn't exactly what I mean by fitting together—there has to be some physical sense to the arrangement that underlies the basic physical mechanism of the ocean machine," says the Chief. So I tell him that is another story that would have to wait till another time, if we happen to meet again later. For the moment it is really just an exercise in showing how they are spatially related.

I sketch Figures 9.1 to 9.3 and point out that two rows of blocks are arranged across the middle of the North Atlantic. The one marked A is the same as the one in Figure 7.3, but with fewer interfaces for simplicity. Another block, A', is drawn in the northern row just to the north of A. (It is not shown in Figure 7.3.) The difference between these two blocks is that A has two interfaces, whereas A' has only one. The upper interface, 1, therefore cuts the ocean surface close to their bounding latitude. We can speak of this line of cutting the surface as the "outcrop" of interface 1. The westernmost block in each row is detached because it is a special case.

Let us now imagine that just over each of these blocks a thin, wind-driven layer (invisible in the figures) is pumping water down (as shown by the vertical arrows) into the top of each block. Rules 1, 2, and 3 apply to the submerged layers in each block: layer c only in the northern row, and layers b and c in the southern row. Rule 3W (see Chapter 8) applies to top layer b of the northern row and to top layer a in the southern row. The interfaces are drawn so that they are smoothly continuous from block to block, as they would be if the jigsaw puzzle were assembled correctly. We do not want any cracks in the interfaces at the joints of the blocks.

You can see that in general both interfaces slope downward toward the west except in the detached, westernmost blocks. They are arranged differently in the north-south direction, however, because interface 1 outcrops near the latitude where the two blocks join, leaving layer b exposed to the wind pumping in the northern row, where interface 2 is deeper.

Let us now assume that the deepest layer, c, is at rest, or moving very slowly. Then the flow in blocks A and A' is similar to that described in Chapter 7 (Figure 7.4.) Interface 2 gets deeper in both blocks as we move westward along the rows.

Block E' at the eastern extremity of the northern row requires special attention. Because it lies right up against the coast of Europe-Africa, no eastward or westward flow is possible. If Rules 1 to 3 apply, the vanishing of these velocity components requires that all the interfaces be horizontal along the eastern wall. (Of course, the physics of a real eastern coastline may be more complex, and we may need some variances of the rules there, but let us ignore that here.) Figure 9.4 shows an enlargement of the interfaces on all four sides; the interfaces along the eastern wall are dashed. Interface 1 must come up to the surface along the eastern wall because we imagine it to be on the surface, at the outcrop, right up to the wall; once it is at this level at one point on the wall, it has to be the same at all other points on that wall. Interface 2, which we do not envisage as having an outcrop, has the same finite depth all along this eastern wall in both blocks E and E'.

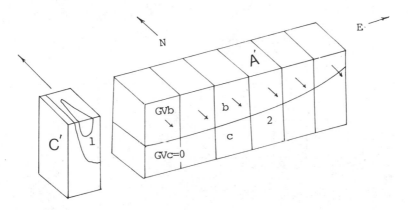

FIGURE 9.1. An assembly of blocks across the central Atlantic, from a Gulf Stream block on the left (west) through six similar but joining blocks across to the coast of Europe-Africa.

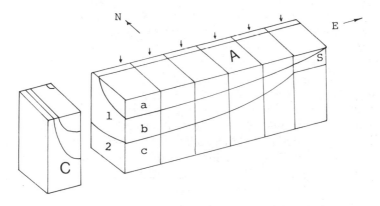

FIGURE 9.2. Another assembly of blocks across the Atlantic just south of the assembly that is shown in Figure 9.1. An outcropping of interface 1 is shown, and in the block farthest to the east a shadow zone S is shown in layer b. East-west arrays of blocks like this pick up more density interfaces at outcrops as we move toward the equator and successively warmer water is forced downward by the Ekman pumping.

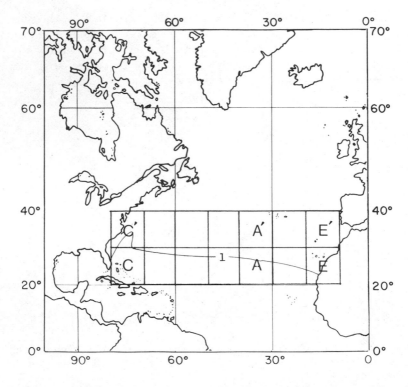

FIGURE 9.3. This figure shows the geographical location of the two assemblies of blocks in Figures 9.1 and 9.2. The contour labeled 1 is the location of the outcrop of a density interface. In the west this warm water is carried northward as a tongue of warm water in the Gulf Stream (Block C').

Now, we can see that the flow of layer b is southward. As columns of it move from the northern to the southern row, they suddenly find themselves covered by a thin layer, a. At this point, Rule 3W for layer b changes to pure Rule 3, and for paths through the southern blocks it must shrink vertically. By Rule 2 the paths of flow in the lower layer must be along lines of constant depth of interface 2, so shrinking cannot occur by a change of depth of this interface, but must occur because interface 1 deepens along the line of flow of layer b. In consequence, the lines of flow of layer b have to be diverted westward away from the eastern wall where interface 1 does not deepen, toward interior regions where interface 1 is deeper. This curving of the lines of flow in layer b toward the southwest leaves an unventilated region of resting water in this layer close to the eastern wall. This stagnant region of layer b has been referred to as the "shadow zone." The eastern boundary of the zone is the eastern wall itself. The western boundary of the shadow zone is depicted as a vertical curtain in Figure 9.4, and the shadow zone is depicted by S. If layer b is not actively moving in the shadow zone, then interface 2 underlying this portion of layer b must be level, and the transport driven by the wind pumping must be confined to upper layer a. This is why the slope of interface 1 is large over the shadow zone. The blocks farther to the west in the southern row have flow in both layers a and b, so both interfaces slope, and their slopes are weaker because they now share the transport due to wind pumping. In the southern row the flow in layer a comes mostly from the accumulated wind pumping of warm water entering by way of the wind pumping south of the outcrop of interface 1. The flow in layer b comes from water subducted at the outcrop, and is introduced from layer b in the northern row. The shadow zone lies south of the dashed curve in Figure 9.5.

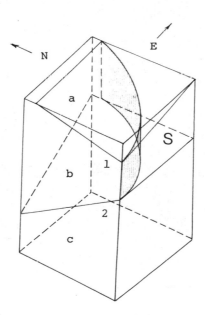

FIGURE 9.4. This is a more detailed picture of Block E, at the easternmost part of the assembly shown in Figure 9.2. We see the outcrop of interface 1, the vertical curved surface delimiting the shadow zone S in layer b, and the flatness of interface 2 under the shadow zone.

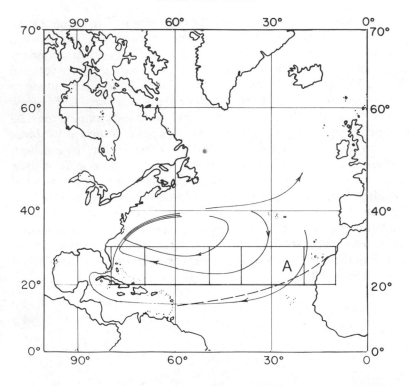

FIGURE 9.5. The geographical extent of a shadow zone with respect to the total circulation streamlines. The shadow zone lies south of the dashed line.

The direction of flow of the GVs in the two moving layers is not the same. Generally, the deeper layer flows somewhat to the right of the upper layer. This change of direction of the GV with depth is our now familiar beta-spiral. In the real ocean, the density is continuous with depth; therefore the spiral is continuous too. Our view depicting the ocean to be made up of a discrete number of layers is an artifice that we use for the sake of simplicity.

Most of the layers shown in Figures 9.1 and 9.2 have a component of flow that is southward or zero. How does the water return to the north? It surely does so in the real ocean because if it did not, the sea level in low latitudes would rise at something like 100 feet a year.

The necessary return flow that goes northward in layers a and b is accomplished by a special regime in the westernmost blocks of each row, Blocks C and C' (Figures 9.1 and 9.2). Rule 1 holds in these two blocks, but Rules 2 and 3 are modified. In the real North Atlantic these blocks represent the Gulf Stream, in which flow returns northward. We can revise Rule 2 to permit a down-the-pressure gradient acceleration in the direction of flow. We modify Rule 3 to allow for the presence of strong relative vorticity, and so we must use Rule 3RV here. In the surface layers of the real Gulf Stream, the relative vorticity of the fast-moving portions of the stream is so great that a barge drifting there would in principle (in the absence of disturbing winds and eddies) turn clockwise, tending to follow the sun. In other words, the earth's rotation is almost canceled out. This is because the strong, northward-flowing, narrow current of the Gulf Stream carries water with smaller absolute spin from low latitudes, and it lags behind the rotation of the earth at higher latitudes. The strong carrying power (*advection* is the technical word) of the Gulf Stream also carries warmer layers northward—warmer than what is characteristic of the surface at the latitudes outside the Stream. We can see this as a thin, warm filament in Block C' riding on top of the strong current associated with the large slope of interface 2 beneath it (Figure 9.2).

THIS IS PROBABLY a good place to say something more about the Gulf Stream, even though we have not visited it during our cruise. Why do the currents that return water to the north appear on the western side of the ocean and not on the eastern side? We will also discover in the Gulf Stream that water does cross isobars, and breaks Rule 2.

Let us suppose, for the sake of argument, that the water moves toward the eastern wall. To accelerate northward along the coast it must get thinner; but if it does, the HPG is eastward and the GV is southward—in the wrong direction. On the other hand, water approaching the western wall (the east coast of North America [Figure 9.5]), wanting to flow northward, finds that as it gets thinner in order to accelerate northward, most of its HPG is directed westward, and this is consistent with a northward GV. That is how we get our northward-accelerated return flows on the western sides of oceans. The relative vorticity is not small in a western boundary current because the velocity is large—several meters a second—and the stream then has to be narrow; narrowness and high velocity go together because of conservation of mass. This leads to a large shear around a locally vertical axis across the Gulf Stream, known as relative vorticity. In the western boundary currents, then, neither Rule 2 nor Rule 3 applies in an unadulterated form.

Figure 9.6a shows four blocks along the western side of the Atlantic Ocean encompassing the Gulf Stream. From left to right in this figure,

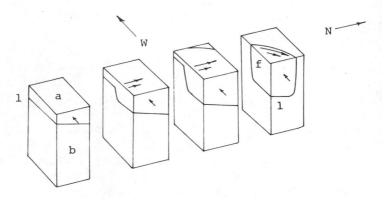

FIGURE 9.6a. Four blocks along the western side of the ocean to show the formation of the Gulf Stream as flow enters from the east into top layer a. The observer is high above the ocean looking toward the northwest. The lower layer b, for extreme simplicity, is taken to be at rest. Starting at the left, southern block, we see water entering in the top layer and accumulating in a narrow northward-flowing stream. Interface 1 slopes down toward the north on the eastern faces of the blocks to ensure inflow, but it slopes sharply up toward the west at the front marked with an f to provide the HPG for the strong northward GV of the Gulf Stream. In the northernmost block, the Stream turns eastward. The relative vorticity of the Gulf Stream is indicated by the difference in length of the arrows on the top of the blocks, the western inshore portion being stronger.

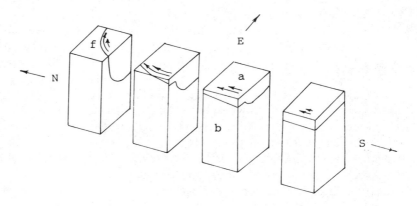

FIGURE 9.6b. This is the same set of Gulf Stream blocks but arranged so as to be seen from the other side (the side facing the United States coast). We see that interface 1 on the western sides of the blocks actually slopes up toward the north, or the opposite slope from that on the eastern faces of the blocks in Figure 9.6. This does not indicate a GV; the northward HPG is used here to accelerate the northward-directed flow of the Gulf Stream—an exception to our three rules.

the span of latitude is 40° beginning at the equator. The view of the blocks is from high above the ocean surface looking toward the northwest. There are only two layers in this illustration; layer a at the top is moving, and layer b is at rest. Thus, for simplicity, we are dividing the ocean up differently here, putting all of our flow in the top layer. Along the eastern faces of the three southern blocks, a westward inflow of layer a water from the interior of the subtropical gyre goes into the Gulf Stream. It enters with no relative vorticity and tries to find its way northward. The large slope of interface 1 down toward the north along the three vertical faces is indicative of an HPG directed southward, and consequently we have a westward GV in layer a, as indicated.

This water is compressed into a narrow northward current, whose high velocity is indicated by the large northward arrows in Figure 9.7. The largest is at the westward side of the current and shows the large shear about a vertical axis and the large relative vorticity (or spin). The depth of layer a on the western faces of these blocks gets smaller toward the north; this supplies the HPG that accelerates the current northward, providing it with high velocity. Figure 9.6b shows the view from the west and the slope of interface 1. It is an example of a locally "downhill" flow of water in the ocean and violates the simple geostrophic ideas that would imply an eastward flow out through the western wall. The large slope (indicated by f) downward across the current toward the east indicates that there is also a strong HPG westward in layer a, and this is geostrophically balanced by the strong northward velocity in the stream. We therefore have a curiously mixed case of balances in the Gulf Stream. Rule 2 applies to the westward HPG and the northward GV, but a Bernoulli acceleration acts between the northward acceleration and the northward HPG. In the rightmost (northernmost) block, I have indicated that the stream begins to flow eastward, but I have not explained it.

Figure 9.7 is designed to show a little more detail in a western boundary current. I have now introduced two moving layers again, a and b, above a resting bottom layer, c, along with two interfaces, 1 and 2. The flow into the blocks from the east is in both upper layers, and both layers move toward the north in the boundary current. The main difference between this figure and Figures 9.6 a and b is the addition of another block at the extreme right that is supposed to lie just east of the fourth block in the north-south line. It shows the extension of the Gulf Stream toward the east as it begins to leave the coast. I have indicated that there is some loss of buoyancy, B, in this block. The block corresponds more or less to block D of Figures 6.2 and 6.7, in

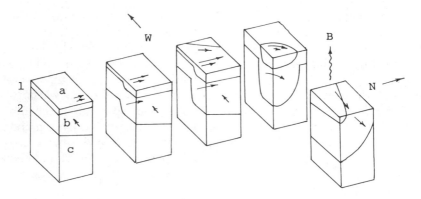

FIGURE 9.7. This picture is an elaboration of Figure 9.6b to show two moving layers and two interfaces. The extra block shows a buoyancy loss B due to rapid cooling and a flow across interface 1.

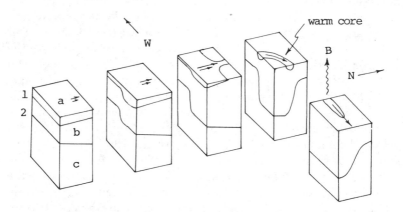

FIGURE 9.8. The difference between this figure and Figure 9.7 is that here we have made interface 1 very shallow, so that it outcrops along the eastern face of the third block from the south, leaving a warm core being advected northward and then eastward in the Gulf Stream. The buoyancy loss B due to cooling wipes this warm core away in the fifth block.

which the geostrophically layered stream widens as a result of the cooling and transfer of water from layer a to layer b.

Figure 9.8 shows how a very thin surface layer, a, which does not reach far north in the interior of the gyre, can be carried farther north in the strong current of the Gulf Stream as an isolated thread of warm water, or as a "warm core." With heat loss at the higher latitude, the volume transport of the warm water in the warm core diminishes and eventually vanishes. In the real Gulf Stream the warm core can be 200 meters thick and 20 kilometers wide.

IT IS GETTING late. The waiters in the cafe are beginning to stack the chairs on the tables. The back of the pages of the Chief's manuscript are covered with my sketches and scrawls. "You'd better keep it," he says, smiling, "for that maintenance manual you're going to write some day. Just remember to mail me a copy."

My plane will leave early the next morning. The Chief is about to go off for a short visit with a lady friend whose beach-front house has been lonely since her husband died. The Chief seems rather jaunty. I have the urge to tell him a great deal more about how we think about the ocean. So far we have only discussed ideas based on observed data. There still is the whole world of theoretical modeling to go into.

Anyway, it is too late at night for that. We walk back to the ship through the dark, wet, quiet alleys. It is time to say goodbye, and I don't expect that we are going to meet again for those other discussions about the machinery of ocean circulation.

Ten / The Luyten Algorithm

TWO YEARS passed before I saw the Chief again. Last summer he came to Woods Hole to retire from the sea for good. He bought himself a small Cape Cod-type house among the cranberry bogs of Hatchville. Maybe he would have liked to have a view of the ocean, but it was beyond his purse. Anyway, he did get himself a small boat to amuse himself with fishing. He did not know many of the people around here, and I suspect he really did not want to know them; he was used to another sort of people and to a world these landsmen did not know.

I used to see him some evenings, sitting on the dock of the Oceanographic Institution—away from the throngs on Water Street—with a fishing pole. Not that he expected to catch anything there, but it was an excuse for sitting alone and looking out over the water as dusk set in. Perhaps it was his way of recapturing those moments at sea when he would emerge from the engine room for a smoke before dark. He would find a sheltered corner on the fantail, behind a winch, for a half-hour's contemplation, before going below to view the nightly movie in the mess—a form of entertainment that he professed to despise.

Or else you could find him at the town dock when the draggers came in after five or six days of trawling on the Great Round Shoal. He was a favorite among the fishermen, offering them useful advice on their problems with machinery, and taking a few lobsters home in return.

My wife's father was sick with his last illness, so she had to spend some weeks out in the Midwest where her parents lived, and I was left adrift. I had just finished mowing the back field and, drenched with sweat and mosquito repellent, was putting the mowing machine back into the toolshed when the Chief put in an appearance. I was surprised but glad to see him, so we grabbed some beers out of the refrigerator and sat down in the shade under the spruce trees, where a fresh southwest breeze cooled us off.

"I'VE BEEN MULLING over those ideas you told me about on the *Atlantis II* since I saw you last," he finally ventures, "and I think that there must be a good deal more about the story than you told me. You oceanographers seem to have some ideas about how the ocean machinery works, and how to test these ideas. But have you ever given

86

them the ultimate test? I mean, have you ever been able to make a model that works? I don't quite know what kind of model, but surely the only way to know whether a machine that is supposed to operate according to certain principles will actually work in practice is to model it. In the old days, the U.S. Patent Office was so bedeviled with crackpot Yankee inventors seeking patents on impractical devices that they used to insist on receiving working models. They don't do that anymore except for perpetual motion machines, but there does seem to be an element of common sense about the proposition."

I can see that the Chief has not lost his punch, and I welcome the opportunity to tell him about some of the theories that have developed since our cruise. A young physicist, Dr. James Luyten, and I had made what we thought was some real progress in modeling the ocean since the last time I encountered the Chief. Even though it might be a little esoteric, I decide to try to explain the model to the Chief.

I first want to describe the kinds of models we work with in physical oceanography. The Chief already understands that the exercises with the blocks of ocean water that we had talked about during our cruise on the *Atlantis II* were not really models at all. They were what we might call "diagnostic studies": they took the density field of the ocean as observed in some 10-degree block of ocean water and determined whether the findings were consistent with certain rules of hydrostatics, geostrophy, and conservation of potential thickness. These studies yielded some idea of how the water must flow at various depths. But we did not start with forcing by wind and heating/cooling, and we did not derive from the rules just what the density field ought to be in the blocks. Therefore our diagnostic exercises did not constitute a model of the ocean.

The first thing that springs to mind that might qualify as a model is a large tank of sea water, rotating like the earth, on which we arrange fans to blow various patterns of wind, with various heating and cooling coils arranged at the water surface in the tank to mimic forced heating and cooling. Such tanks have actually been built, but they have been successful only in a very limited way. Unfortunately, most real fluids are too viscous to obey the three rules we used on a laboratory scale—it takes the grand scale of the ocean itself to make them do so. And even more fatal to the usefulness of rotating tanks is the fact that one cannot arrange them to produce different rotations at different latitudes. Water will not stick to a sphere unless the gravitation of the sphere dominates that of the earth, and we would have to build the tanks out of the superdense material from the core of a white dwarf star, or conduct the experiments in a space station.

Large, numerical computer models that contain the hydrodynamical equations in sufficient detail to mimic the ocean have been constructed. They have the advantage of being sufficiently flexible to set viscosity to zero if desired, and they do incorporate the differential rotation of the earth at various latitudes. The computer models include various elements, such as artificial horizontal mixing, so our three rules can't quite apply. These models generate such masses of tabular data that they are as much a challenge to understand as the ocean itself. Consequently, the numerical results seldom get the detailed study, interpretation, and explanation they deserve. In the hands of some they are a wasteful, idle exercise.

So I tell the Chief what we have been up to and show him some of our theoretical models. They are purely theoretical because we start with the idea that the ocean is made up of three layers, two of which move, separated by two interfaces. We call the top layer the *warm layer*, the next one down the *cool layer*, and the bottom one the *cold layer*. The two interfaces are distinguished by the words "upper" and "lower." Sometimes the upper interface reaches the ocean surface at an outcrop; north of the outcrop there is no warm layer at all: the cool layer is then exposed to the surface, and as you go down in the ocean you pass through only one interface. I make a sketch for the Chief to show him a block of ocean that illustrates these definitions (Figure 10.1).

The Chief inspects it and comments: "I notice that you indicate the horizontal directions by x and y instead of by east and north. That's mathematical to begin with." I tell him that we do that all the time because we have to use some algebra and slightly higher mathematics to work out our deductions or predictions from the model. At one time I had wondered why we chose the symbols x and y—the conventions used in high-school analytical geometry (the Chief said he had had some). Then I discovered that Descartes, who had invented them centuries ago, probably picked them up from old charts that in his time used an x to depict the Cross, pointing in the direction of Jerusalem, and a T or fleur-de-lis pointing toward the north.

Let us suppose that this block represents a larger hunk of the ocean than just 10 degrees. We can, for example, place the eastern coast of the ocean along the side of the box on the east, which means no water can flow through there, so that the slopes of the interfaces in the north-south direction along the eastern side of the ocean have to be zero. Elsewhere the interfaces can slope and geostrophic velocities can occur; but we now want to be able to deduce what these slopes are, not just get them from observation. If we can work out a way to do this mathematically, and if we find that the result corresponds in some respects to observations, then we can speak, in a sense, of having computed a model that looks like the real ocean; that gives us some confidence in the rules we used to get the results. So that is what we mean by a model: we want to use it to compute the dependence of the two interfacial depths h and D (see Figure 10.1) for all values of x and y in the region under consideration. The main input is the vertical flux of water w_e from the wind-driven Ekman layer on the top (not shown in the figure), and the interfacial flux across the upper interface w_s, which represents the net heating or cooling of the ocean locally. These two driving agents, w_e and w_s, should be known beforehand as functions of x and y. We want to compute h and D as functions of x and y.

It turns out that this computation is not easy to do with paper and pencil and a table of known functions such as exponentials and sines and cosines. We will need the help of a microcomputer to serve as a sort of chauffeur to drive our model for us. But unlike the huge computer models, we are not at its mercy. We can follow in general terms what it is doing at every moment, as it develops the fields of h and D, and we will use it as a convenient device that can do our detailed summing and multiplying for us.

The Chief scratches his chin and comments that this sounds like an

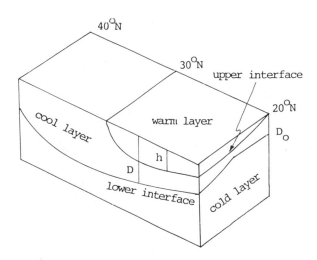

FIGURE 10.1. The sketch of the layered model that was made to run on a personal computer, defining some of the terms. The upper interface is here shown as outcropping at 30°N and along the eastern wall; the lower interface has a fixed constant depth along the eastern wall.

automatic pilot to him. I agree, and hasten on with the details. Let's suppose that our model ocean is driven by an Ekman transport pumping down from the wind-mixed layer that has a dependence on latitude, as shown in Figure 10.2. You will notice that the downward velocity w_e is largest at mid-latitude at about 30°N and tails off to zero both to the south and north. This corresponds to what would happen in the region between the westward-blowing Trades and the eastward-blowing winds that prevail at higher latitudes. These winds and the form of w_e are envisaged to be independent of longitude. Now, I will add two important theoretical rules, Rules 4 and 5, to our previous three, and they will enable us to construct the geographic dependence of both h and D over the interior of the ocean.

Rule 4, the *total transport rule*, was first written down by the oceanographer Harald Sverdrup in 1947. He did not write it in the form I will use, but that is a small matter. My purpose here is to clarify the main ideas of the model, and so we will ignore various dimensional and geometrical factors. We will assume that there is no motion in the deep cold layer, so that only the upper and lower layers move, and we will assume that the density contrast across each of the interfaces is

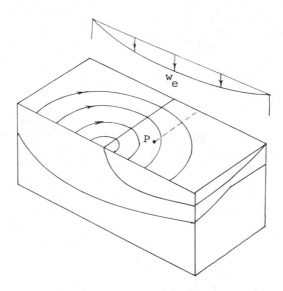

FIGURE 10.2. The same model as in Figure 10.2, but this time showing how we specify the Ekman pumping as a function of latitude, and showing the resulting streamlines of the Sverdrup transport function $h^2 + D^2$.

the same. Then the geostrophic velocity in the northward direction (GVN), averaged over the depth D, is simply,

$$GVN \sim w_e/D.$$

The GVN depends on the local intensity of Ekman pumping w_e. Inasmuch as w_e is negative over the center of our gyre, the GV averaged over depth D (which we can call the *Sverdrup velocity*, or SV) is directed southward, as shown by the arrows in Figure 10.2. At higher and lower latitudes, this Sverdrup velocity is smaller, and since we have blocked off the eastern wall from flow, mass conservation requires that there be an eastward inflow from the west in the northern half of the gyre and a westward flow in the southern half of the gyre, as shown in the figure. In this way, conservation of mass requires that the Sverdrup velocity has eastward- and westward-directed components. Both components of the Sverdrup velocity, SVE (eastward) and SVN (northward), can be computed from the relations,

$$SVN \sim (1/D) \frac{\partial}{\partial x} (h^2 + D^2); SVE \sim -(1/D) \frac{\partial}{\partial y} (h^2 + D^2),$$

where the quantity $h^2 + D^2$ at any point is computed from the expression,

$$h^2 + D^2 \sim 2 \int_X^0 w_e \, dx + h_0^2 + D_0^2,$$

where h_0 and d_0 are the constant values of h and D that we have assumed along the eastern wall. The contours of this function, $h^2 + D^2$ are drawn as the curves in Figure 10.2 and show the direction of the Sverdrup velocities in the gyre.

This method of finding the field of the combination $h^2 + D^2$ is called Rule 4. It is easy to compute.

"Well," says the Chief, "that's O.K. as far as it goes, but it doesn't give us either h or D separately, just the sum of the squares. One couldn't make much use of the Pythagorean theorem of trigonometry if all you were given to work with was the hypotenuse. I suppose that the Rule 5 that you promised is going to give us one of the sides so that we can get the h and D individually." Our Chief, you see, is right on the mark.

Rule 5 is the Veronis characteristic equation. It is a differential equation for D driven by buoyancy flux w_s.

$$dD/ds \sim -h \, w_s/D.$$

On the left-hand side we have the unknown rate of change of the depth D as we proceed along a curve whose arc length is s. We don't know ahead of time just what the shape of this curve is going to be, but a point moving along it moves with an eastward and northward *characteristic* velocity (CVE and CVN, respectively), given by the following expressions:

$$CVE = SVE + RW; CVN = SVN.$$

You will notice that the northward characteristic velocity (CVN) is nothing more than the SVN of Rule 4; however, the eastward characteristic velocity (CVE) is made up of two terms, the SVE and the RW (Rossby wave velocity), which is of the form,

$$RW \sim -h(D - h)/D,$$

where RW stands for Rossby wave velocity.

"That Rossby wave velocity, RW, is perplexing," interrupts the Chief. "We're discussing steady solutions without waves, aren't we?" I reply that that is what we are indeed doing, and that the RW term is just the mathematical expression for the velocity of a Rossby wave. It does not seem wise to digress into the details of Rossby waves at the moment, although we could do so easily enough. In fact, they are manifestations of our three rules in which northward-moving columns of water actually do stretch in length and southward-moving ones shrink. As they do so the HPG distribution changes, and so do the GVs. The result is that the form of the h and D interfacial depths progresses toward the west in time. The term RW happens to tell us what the speed of such a westward progression would be. But we are discussing a steady state in which columns maintain their heights steadily in time through the agency of the Ekman pumping and interfacial mass fluxes. So let's get back to the mainstream.

Because RW is always negative (for positive h and $D - h$) it tends to make the point that generates the characteristic curve drift off the contours of $h^2 + D^2$ toward the west. When either h or $D - h$ is small, this rate of westward drift is small, and it is at a maximum when both are the same. Because this extra westward term is identical with the phase speed of a nondispersive Rossby wave, we denote it by RW. As a result of RW, the characteristic curves will differ from the contours of $h^2 + D^2$ whenever both h and $D - h$ are not zero.

A form of this equation was first written down by Yale professor George Veronis in 1978. George is one of my oldest friends and colleagues. We have written many papers together over thirty years' time, in the course of which he taught me what little mathematics I know. But when he discovered this equation, he did not, I believe, recognize the simple interpretation in terms of Sverdrup velocities and Rossby waves given above. George is one of the pioneers in macrocomputer programming, and when I first met him in the early 1950s he was working on von Neumann's computer at the Institute for Advanced Study in Princeton, New Jersey. In 1978 the microcomputer craze had not yet caught up with us. So, to my knowledge, Jim Luyten was the first to recognize this interpretation and the first to succeed in combining Rules 4 and 5 into a useful algorithm for use on a microcomputer to compute and display the fields of h and D with enough flexibility to find the new physical phenomena that lay hidden in the model. It is for this reason that I like to speak of this powerful combined use of Rules 4 and 5 as the *Luyten algorithm.*

Now let us see how we can employ the Luyten algorithm to compute the fields of h and D for us. The first thing to notice is that in Figure 10.2, where the flow along the contours of $h^2 + D^2$ enters the block from the west, there is no upper layer, so that RW = 0. Thus in the region north of 25°N, where h is zero, the characteristic velocities CVE and CVN (which determine how our point moves) are simply SVE and SVN, and the point sticks to the Sverdrup transport contours. Once we cross the latitude 25°N, which in this example has been chosen as the place where the temperature of the water being pumped down from the Ekman wind-driven layer with velocity w_e is suddenly switched from cool to warm, the warm layer begins to form on top of the cool layer and h is no longer zero; neither is RW. Therefore, the moving point that describes the characteristic curve begins to drift westward across the $h^2 + D^2$ contours.

I sketch some characteristic curves described by points moving under control of the Luyten algorithm for the Chief (Figure 10.3). In the northern region, where $h = 0$, they follow the $h^2 + D^2$ contours; but in the southern region, which has a warm layer and where h is not zero, they drift toward the west up onto higher values of the Sverdrup transport. This drift toward the west tends to leave a large region in the southeast of the model that is unreachable by characteristic curves coming in from the northwest. This is easy to see: look at the characteristic curves near entry point A, for example: they are swept so abruptly away from the eastern wall that they leave a "shadow zone" along that wall in the southern half of the model. This shadow zone was discovered by Luyten, Pedlosky, and me and reported on in a 1983 paper called "The Ventilated Thermocline;" but the Luyten algorithm is much more general and clearer.

Characteristic curves cannot enter from the west at lower latitudes where CVE is negative. In the model we have set up here, they cannot enter from the north or south because along these boundaries CVN = 0.

The only other place at which characteristics might enter is in the southern warm-water region along the eastern wall, where CVE is weakly negative. (It is zero exactly at the wall, but negative just within it, so there is a special argument for making the first step in the Luyten algorithm, into which we cannot go here.) In Figure 10.3 I drew one such characteristic emanating from point B on the eastern wall and passing southwestward through the shadow zone.

All that remains to be done is to compute the changes of D along the characteristic curve according to Rule 5. The value of h can therefore be derived from Rule 4, and thus we have succeeded in getting the fields of h and D over every part of the model that the characteristics can reach.

"The notion of these characteristic curves is something new to me," says the Chief at last. "I can see that they tend to move with the average velocity of the fluid, but not completely so because of that RW part. So they are not the curves along which the water itself flows in either layer. They might even enter the model at places where water was flowing out of the model in one of the layers. At first they seemed

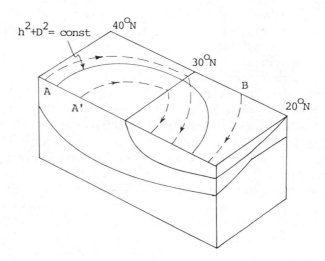

FIGURE 10.3. The same model as in Figure 10.2, but this time showing the paths of the characteristics (dashed lines), some emanating from the west and some from the east.

rather mysterious to me, but then I realized that they are really the curves along which information about boundary conditions flows: they show how our specification of conditions around the box affects conditions within the box. Just take that characteristic curve in Figure 10.3 that starts at boundary point A', for example. It remembers the depth D that it originally had at A as the computation carries it along the characteristic, but it always keeps track of the w_e forcing and eastern-wall boundary condition through Rule 4. According to Rule 5, the characteristic doesn't change its value of D from what it was at A until it reaches 30°N, because along this northern portion of the path $h = w_s = 0$ and $dD/ds = 0$ by Rule 5. Once the characteristic passes point A', where warm water begins to be showered on the top by w_e, h is no longer zero. According to Rule 5, the depth D (with positive w_s) begins to diminish, and h, by Rule 4, increases along the characteristic curve. The manner in which the Luyten algorithm permits us to calculate h and D along characteristics gives us a deeper understanding of how the ocean model feels and responds to the values of h and D specified on the boundaries, and which points on the boundaries determine the interior points. That's a fair distance toward understanding, and I see why you are so enthusiastic about the algorithm."

THE WIND has gone down under the spruce trees and the mosquitoes are getting annoying. We decide to truncate our discussion, but before the Chief leaves he has a few more words to say: "It occurs to me that this algorithm sheds a lot of light on the old debate about whether winds or heating and cooling drive ocean currents. If you just look at Rules 4 and 5 and how they work together in the algorithm, you'll see that in a sense the wind steers the information flow about h and D (because the CVs contain wind-determined SVs, with an assist from the RW, which of course contains the local accumulated information about h and D), whereas you might say the heating and cooling, as expressed by the w_s on the right-hand side of Rule 5 change the information. On the other hand, as a lifelong engineer accustomed to concepts of energy, power, and work, I am a little uncomfortable to be spouting about information flow like some kind of communications expert."

Eleven / A Model on a Personal Computer

ON THANKSGIVING DAY 1984 I bought myself a personal computer. For the first time I did not need a programmer as a liaison between me and the computing power that the modern age has brought to us. It was easy to learn how to program it in BASIC and to get excellent graphics with the color card. I happened to buy a NEC-APC III with color card and spinwriter, but most of the programs that Jim Luyten and I have written are compatible with an IBM/PC. The marvelous thing about all this was that it freed me from the tiresome restrictions of analysis, and we were able to solve problems that we weren't able to work on before. During the spring of 1985 Jim and I started to use the computer to develop models of the sort that I described to the Chief in the previous chapter.

To do anything new in science, I have to become totally preoccupied with a problem—the kind of preoccupation that other people seem to be able to summon up when they are confronted by a lawsuit or a serious illness. The problem simply has to be in the forefront of my mind all the time—at night, on weekends, while I am cutting the grass or painting the bathroom. I found in Jim a person who could also give the problem of ocean circulation his full attention and enthusiasm, and he had mathematical skills and insight far superior to my own. It has been a delightful collaboration. One afternoon he happened to write down his algorithm on the blackboard, and from then on we were able to make further headway in our modeling. The computer's ability to integrate along the characteristics, as well as its provocative color displays of results, helped to reveal many phenomena that are new to oceanographic theory, and I think they will occupy a place in future textbooks. We built up a small library of programs, each designed to illustrate some particular feature of the ocean's theoretical response to wind and heating/cooling combined.

ONE JULY morning the Chief drops into my office. I slip a disk into the computer and show him how we actually run the example of a subtropical gyre driven both by w_e and w_s. The form of these two driving functions that we use is shown in Figure 11.1. The range of latitude is from 20°N to 40°N, and we place the maximum downward (negative) value of w_e of -3 cm/day at 30°N. North of this latitude the water being forced down by the wind is cool, and south of it, it is warm. The heating function is represented by a flux w_s across the upper interface from the cool to the warm layer, and the maximum of 3 cm/day is placed at 25°N. In this case, w_s is positive, corresponding to a fixed rate of heating at the upper interface. Our choice for the depth of the two interfaces is constant along the eastern wall: $h_0 = 0; D_0 = 150$ m.

By Rule 4 we can immediately calculate the field of the total Sverdrup transport, $h^2 + D^2$, as shown in Figure 11.2, and of course we know SVE and SVN from this figure as well. Figures 11.2 through

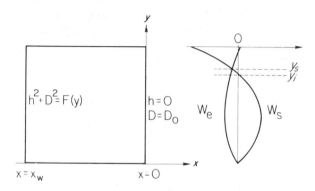

FIGURE 11.1. The Ekman pumping w_e and the interfacial flux w_s that were used in our computed example as functions of latitude in the model. (From Luyton and Stommel, "Gyres Driven by Combined Wind and Buoyancy Flux," *Journal of Physical Oceanography* [Sept 1986]. Reproduced courtesy of American Meteorological Society.)

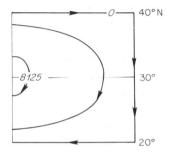

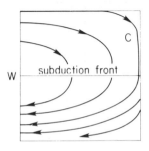

FIGURE 11.2. The total transport computed.

FIGURE 11.3. The subduction front and chracteristic paths.

11.5 are computer graphics. They look much prettier on the screen because they are in color, and you can watch them being drawn.

Once we have entered the numbers into the computer, I ask the Chief to press the RUN key, and the main display begins. A characteristic curve forms at the top left of the plot, which is a plan diagram of the model. Looking down, north is at the top and east is on the right, as shown in Figure 11.3. The first characteristic curve, indicated by C in the figure, steers itself with the Sverdrup velocity over toward the east wall and then down south, following the $h^2 + D^2$ contours. Since $h = 0$ in this region, it really follows the D-contours. When it reaches the latitude (here 30°N, where the w_e forces warm water down and h is no longer zero), it starts to feel the influence of the Rossby wave velocity and begins to drift off the $h^2 + D^2$ contours toward the west, avoiding the southeast corner of the box. According to Rule 5, it also begins to compute shallower values of D along its path, until it exits

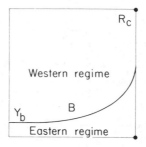

FIGURE 11.4. The western and eastern regimes, their boundary B and the location of the Rossby repellers R_c.

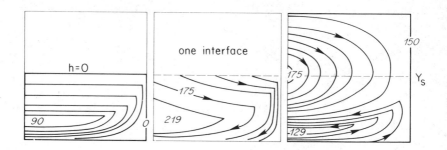

FIGURE 11.5. Computed contours of h, $h + D$, and D. The geostrophic velocities of the lower layer are shown by the arrows on the right panel, those of the upper layer on the center panel.

from the box on the western wall somewhere south of 30°N. Since in its drift across the $h^2 + D^2$ contours it drifts toward higher values of $h^2 + D^2$ and D itself decreases, h increases all the time along the characteristic. So we now have the values of h and D along one of the characteristic curves of the problem.

To get a reading of h and D individually over the entire box, we obviously need to calculate only the characteristics that fill up the box entirely. For example, the next characteristics below C that start from the western boundary begin with a larger D; because in the northern half they flow along lines of constant $h^2 + D^2$, they do not reach over so far to the east before striking the latitude 30°N and subducting under the warm top layer. They complete their cycle by exiting from the box north of the exit point of characteristic curve C. After all these characteristic curves are displayed on the screen, the computer goes over to 20°N on the eastern wall and starts to draw characteristic curves from there, beginning with the boundary values h_0 and D_0. As they drift westward steered by a CVE in which both terms are negative, D decreases along them as well. The computer continues to draw the curves at starting points at successively higher latitudes until it reaches the latitude of subduction at 30°N; then it stops plotting. As we can see from Figure 11.4, the ocean model is thereby divided into two regimes: one with characteristics that originate in the west, the other with characteristics that originate in the east. The former carry information about boundary conditions specified on the western boundary; the others do not. As we will see when we look more deeply into the physics of these two regimes, they are very different in other respects as well. In the northeast corner, a point marked R_o is unique in that no characteristic can approach it, and accordingly we call it a *Rossby repeller*. In certain configurations a Rossby repeller can occupy a mid-ocean position, and that is interesting.

"I hope there aren't any Rossby attractors in the real ocean," comments the Chief, "for then I'd have to believe in the Bermuda Triangle."

We arrange the graphics so that they can display the h or D as it develops along the characteristic curve; we do this by changing the color of the curve at various intervals of h or D. When all the characteristics have been drawn, the colors stand out clearly and their boundaries define the contours of equal values of h and D. That way we do not have to do a lot of interpolation and contour plotting. I cannot show these colors very easily in a book so I draw, by hand, the color boundaries, and show them in Figure 11.5 as contours of h, $h + D$, and D.

The contours of h and D are the depths in meters of the upper and lower interfaces. The contours of D also show the direction of flow in the lower layer by little arrows computed geostrophically. The geostrophic velocity in the upper layer is governed by the horizontal pressure gradient there, and for this purpose we need $h + D$. The little arrows on the $h + D$ contours show the direction of flow in the upper layer.

An inspection of these diagrams reveals that the shape of the gyres of flow in the two layers individually is considerably different from that shown by the total Sverdrup transport in Figure 11.2. For example, the upper layer occupies a pool in the lower latitude half of the model, with zero depth around the northern, eastern, and southern boundaries, but with a maximum depth of about 90 meters on the western boundary near 23°N. Its circulation (as shown by the contours of $h + D$) is skewed, so that in the western regime, outside of the shadow zone, the water in the warm layer moves from shallow to deeper parts of the pool, in a generally east-by-southeast direction. This movement carries cool water from the cool layer into the warm layer across the sloping upper interface, even though the vertical component of velocity is downward there. In the eastern regime the flow in the upper layer more nearly parallels the contours of equal depth of the upper interface, and the flow from cool to warm layer across the upper interface is provided by a truly upward vertical velocity there, associated with the vertical stretching of columns of cool layer water that are, in the eastern regime, moving poleward. The contours of D, the depth of the lower interface, give the direction of the cool-layer flow. There are two gyres in this layer. To the north is a great clockwise gyre that brings water from the western boundary and subducts it under the front of warm-layer water that starts at 30°N. This gyre does not extend as far south as the boundary between the western and eastern regimes, which is straddled by another gyre in the cool layer that carries water around in a counterclockwise direction—out of the western boundary at low latitudes, then north until it hits the regime boundary, and then it turns sharply southwestward to snuggle just south of the northern gyre and eventually flow out through the western boundary, again at about 23° to 25°N. This gyre is clearly associated with the driving due to heating and cooling. It was first portrayed to the world of science by our little personal computer screen. Needless to say, it was a great thrill to Jim and me when we got the first version of this graphics program running and discovered the many unsuspected features it showed.

"It really is quite nice," says the Chief. "I suppose that the place in this model that corresponds to the location in the North Atlantic

Ocean that we were surveying on the *Atlantis II* back in 1981 is somewhere toward the east in the western regime, just north of the boundary of the shadow zone, but south of the subduction front. I notice that in this whole area the direction of flow in the cool layer is considerably to the right of that in the upper layer—your 'beta-spiral,' so to speak. Lucky that you didn't stumble into the shadow zone a little closer to the Cape Verdes and miss seeing a spiral entirely. Then you might have given up on this stuff before you should have."

Before leaving my office, the Chief comments: "I must say that your idea of a working model is novel to me—I generally think of something made of cast iron and brass, not a theoretical algorithm running on a little home computer. However, I'll take these sketches with me and give them some thought." So the Chief leaves, and I don't see him again for a few weeks.

Twelve / The Chief Gives Me a Lecture

ONE DAY the Chief and I find ourselves in the port engineer's office overlooking the dock, waiting for a scheduled radio contact with one of the ships. The Chief tells me that he has been thinking about the demonstration of the Luyten algorithm and the computer-guided model of the two moving ocean layers that I had shown him. "I'm a sliderule man myself," he says, "and so I can't say much about the computer except to admire the display. Where'd you get that idea of using color for the contours of h and D?"

So I tell him that story. Back in 1962 I met a computer genius named Ed Fredkin who had off-hour access to one of the early PDP-1 computers of the Digital Equipment Company, at a time when it was first starting up in an old warehouse in Maynard, Massachusetts. We spent many nights there trying to put together a computer-generated atlas of hydrographic station data. To avoid getting mixed up with a contour-plotting program, we showed temperatures, etc., by using dots of different colors. Ed was twenty-five years ahead of the times, and his proposals to the National Oceanic and Atmospheric Administration to set them up with what we called a "living atlas" were turned down by the comparatively unimaginative bureaucrats there at the time. So he set up his own company to design programs and devices for the Air Force, and quickly he made millions and became a professor at M.I.T., to boot. On the whole, however, there really have not been many such checks to what I have wanted to do in science. Over the years, the Office of Naval Research and later the National Science Foundation have supported my research very generously. It really is a kind of miracle, when you come to think of it, that an enthusiastic amateur scientist like me can get any financial support at all to study the ocean in the abstract kind of way that I do. What a wonderful century it is to live in. My life has been a good one.

The Chief had of course made his own acquaintance with long-dead bureaucracies, so he does not say anything else along these lines as he went on: "As I was thinking over what you showed me, it occurred to me that here is a little model ocean in which you actually know that the shape of the density interfaces varies with latitude and longitude much more precisely than you could ever hope to measure them in the

102

real ocean. So why not examine it by the block methods we talked about during our cruise together? I know it is rather redundant, because the results are supposedly guaranteed to work out; but going through the exercise might reveal interesting things about the physics. As I see it, the Luyten algorithm has the same physics as old Rules 1, 2, and 3; it is just a way to use them to calculate the density surface configuration. So once we have its results in a particular case, we ought to be able to test it with the diagnostic method and maybe learn something on the way."

We find some scrap paper in the form of old, canceled nautical charts on the drawing board. The Chief turns over one of them and sketches Figure 12.1, all along explaining what he is doing: "Consider one of the characteristics that comes out of the western wall. Now, I'm going to put one of our famous diagnostic blocks, B1, astride the curve north of 30°N in the region where there is no warm layer. Incidentally, I can make this box a lot smaller than the ones you had to use in the real ocean because here the data are so good that I don't need horizontal size to get a good estimate of the slope of the interface. There is just one interface, the lower one, and, looking back at Figure 11.5, which showed the contours of D, I can fit a plane locally to it, as in this figure [Figure 12.2]. The slope of the interface is such that the northwest and southeast corners of the box have the same level (they lie

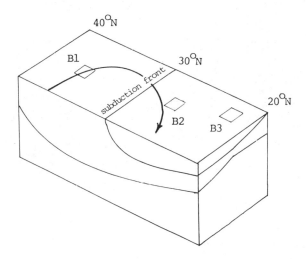

FIGURE 12.1. The Chief's sketch map showing the location of the regions in the model where he made his diagnostic studies.

along curve C, which in this region coincides with a D-contour). The depth D is greatest on the southwest corner and least on the northeast corner.

"According to Rule 1, if there is no HPG in the bottom cold layer beneath the interface at D, then there is a northeastward-directed HPG in the cool layer above it. According to Rule 2, there is then no GV in the bottom layer, and the GV in the cool layer above the interface is directed toward the southeast corner, along the D-contours and the characteristic curve. So far so good." The Chief took a short pause, then continued his little lecture.

"Now we come to Rule 3. Let us consider a column of cool water that enters the block along the characteristic at the northwestern corner. It moves with a GV independent of depth, so it always stands up straight. It therefore moves along a vertical, cylindrical surface generated by the characteristic curve C. This cylindrical surface is shown in Figure 12.2 by the dark area, and also on the right. Its bottom is

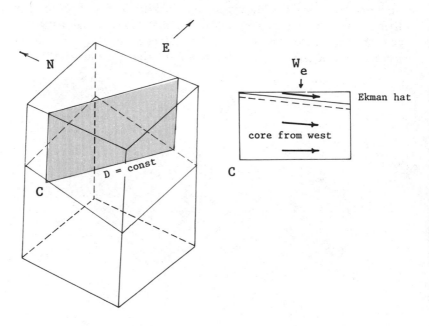

FIGURE 12.2. Block B1 and the vertical cylindrical surface (approximated in the small box by a vertical plane element) through D = constant, with the velocities shown on the right, indicating the shrinking of the columns by Rule 3 and the accumulation of the Ekman hat from the Ekman pumping.

tied to the constant D-contour, and the top is the level sea surface, so it can't change its height. But here we run into trouble with Rule 3, which says that since it has a southward component of motion it must shrink vertically. This was the kind of contingency that the modified Rule 3W is supposed to take care of: the original column does in fact shrink, along the sloping line S, and it has added to its top just enough water from the Ekman pumping w_e to keep the total height constant. So in this region, columns of cool water have Ekman hats added to their tops. They move southward at just the right speed so that the fixed rate of Ekman pumping can keep them at constant height. Rather marvelous, I think.

"It also occurred to me that this hatting business doesn't start in Block B1, but actually starts at the very beginning of the character-istic curve at the western entry point, so I ought to place the hat brim farther down, along the dashed line, to show that there already was some Ekman hat when the water came into the box.

"When the column finally reaches the latitude 30°N, where the tem-perature of the Ekman pumping switches from cool hat to warm hat, the cool column is made up of a certain fraction of cool Ekman hat and original cool water from the west—so there is a good deal more water circulating along D-contours in this model than is supplied by the Ek-man pumping in one pass through the model. This is a kind of ampli-fication which seems a little surprising to me."—"We call it *recircu-lation*," I interrupt.

"Now I come to the hard part," continues the Chief. "Once we get under the subduction front, our cool column, Ekman hat and all, doesn't follow the characteristic curve anymore. It has to follow a D-contour just as it did before by Rules 1 and 2. From what your model shows, I chose another block, B2, south of the characteristic curve but on the same D-contour so I can continue to follow the same column. When I fitted planes to the interfacial depths h and D in this block, it looked something like this [Figure 12.3]. The upper h-interface slopes upward toward the north; the lower D-interface slopes down toward the northwest.

"The cool column is now confined between the two interfaces, and it moves along the contour of constant D, with a speed given by Rules 1 and 2. Since it has a southward component of flow, it must be expected, by Rule 3, to shrink further in the vertical, or the total, column—that is, including the Ekman hat it had just at subduction. Of course, no more Ekman hat is added to this cool column after it subducts. If there is no interfacial flux across the upper h-interface, then this is a pretty strong restraint on the column's path, but the model ought to find it.

When I looked at the numbers, however, I found that the thickness of the cool layer was diminishing at too fast a rate to accommodate Rule 3, and inasmuch as I could not possibly resort to Rule 3W in this case, because the layer is not in contact with the Ekman pumping. However, it could be due to Rule 3B. The implication is therefore that the cool column in Block B2 is losing its top through the h-interface at a rate consistent with the imposed interfacial flux w_s from the cool layer to the warm layer. In other words, the top of the cool column is getting cut off, and it passes into the warm layer, through heating, where it puts a kind of buoyancy boot on the warm column—and I do want to say more about the warm layer later. There is another remarkable thing about this block. Since the cool column can't tilt, it moves along a vertical cylindrical surface passing through the D-contour. The flow of the cool layer is horizontal at the depth D-contour, and since there is some shrinking according to Rule 3, the upper portions of the cool column must be descending slowly, so the vertical velocity near the

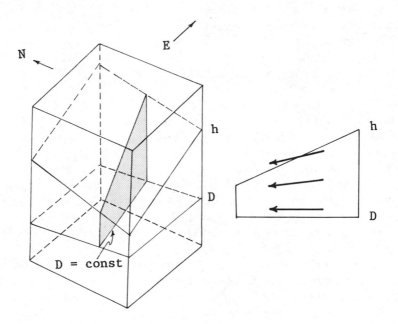

FIGURE 12.3. Block B2 in the western regime and the vertical cylindrical surface through D = constant within the lower layer. To the right the Chief showed how the columns in the lower layer, though shrinking according to Rule 3, could nevertheless transfer water upward across the sloping interface at depth h.

top of the cool column must be downward. The forced w_s at the h-interface is positive for heating, however, from cool to warm layer. How do we reconcile a negative vertical velocity in the cool layer at the level of the h-interface with a positive w_s there? The block model shows it clearly: the slope of the intersection of the h-interface with the vertical cylindrical surface along which the cool columns move is so great that, even with downward shrinking, cool water flows across the interface into the warm layer. The quantity w_s is not a vertical velocity; it is a flux from one layer to another across the interface, and it can be positive even in the presence of a negative vertical velocity if the interfacial slope is great enough." I interrupt again, probably for the shameful reason that my ego is getting involved and I want the Chief to know that we had thought these things out before he did: "We have called that kind of situation *indirect buoyancy convection.*"— "Nice name," he says, with a knowing grin.

Then he picked up his pencil again and went on: "The next thing that I thought about was what was happening in the warm layer in this same box. So I redrew the interfacial depths for Block B2 just as before, and then plotted the direction of the sum $h + D$ from the contour map you gave me [Figure 12.4]. The columns of warm water have to move in the vertical plane that passes through the intersection of the interfacial depth h and the contour of $h + D$, as shown by this sketch. Tracing the $h + D$-contour back, I find that it comes from the western wall, just south of the subduction latitude at 30°N, so that originally it must have contained a rather short column of warm water from outside the model. As it moves toward the southeast, this column actually gets taller, but of course by Rule 3 it would be expected to get shorter. Indeed, that portion of the column that originally came from the west does get shorter: it occupies an ever-diminishing thickness of fluid in the middle of the warm layer, on top of which is a growing warm Ekman hat. On the bottom of it, water heated at interface-h is added (buoyancy boots), and it passes up through the top interface to join the bottom of the warm-water columns moving southward. The slope of h downward in the direction of flow is so strong that the water can pass from cool to warm layer even though it is moving downward. This is the other part of the indirect buoyancy convection.

"Finally, I decided to have a look at both layers in a block of water inside the shadow zone, where the characteristics emanate from the eastern wall. I chose Block B3. This box, as shown in Figure 12.5, seems very interesting to me. Notice that its characteristics come from the eastern wall only, and that therefore it doesn't know anything at all about conditions on the western wall. Nevertheless, it draws water from the western wall in both layers, and drives it back into the western wall, producing a sizable recirculation in this way. Whatever the dynamics of currents to the west of the western wall, they don't seem to have much chance to object."

I tell the Chief that he would have made one damn good oceanographer, and he replies that he has preferred being a good marine engineer, with real steam engines and diesels to work with rather than imaginary machines on the scope of a puny computer. He then goes on with his lecture.

"An astonishing thing found in this region is that the contours of h, $h + D$, and D are all parallel. According to Rules 1 and 2, this means that water columns in both layers move in the same vertical plane, as I draw it here [Figure 12.5]. The columns in the cool layer expand as they go northward, and since they have to move horizontally at the bottom, they have a vertical component of velocity near the h-inter-

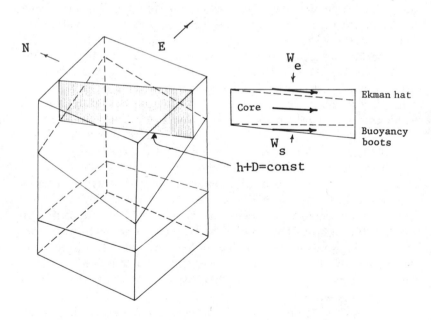

FIGURE 12.4. Block B2 in the western regime and the vertical cylindrical surface through $h + D$ = constant, with the original column's core shrinking according to Rule 3, but with addition of Ekman hats at the top and an accumulation of buoyancy boots from the upward interfacial flux across the interface at depth h.

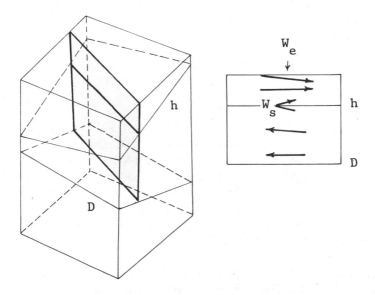

FIGURE 12.5. Block B3 in the eastern regime and the vertical cylindrical surface through all three contours of h, $h + D$, and D simultaneously (possible only in this eastern regime). On the right we see the shrinking of the original column of water in the upper layer according to Rule 3, with the addition of Ekman hat and buoyancy boots, but with the columns of northward-flowing water in the lower layer expanding and producing the needed upward interfacial flux across the level interface h.

face. But it happens to be flat, too, where it intersects the cylindrical surface, so the top of the columns in the lower cool layer marching northward have to punch their way through the h-interface. That is what they are supposed to do to produce the upward w_s prescribed there. In doing so, they put buoyancy boots onto the columns of warm water that are moving southward in the upper warm layer. These columns shrink vertically so quickly that the buoyancy boots don't help them enough, but they also get help from the Ekman pumping downward from the top, w_e, and this turns out to be just what they needed. They need help from Rules 3W and 3B working together.

"This is as far as I've gotten trying to make a diagnostic study, as you call it, of the model."

Thirteen / Complications

WHEN WE MEET again, a few weeks later, the Chief apologizes for monopolizing the conversation we had had in the port engineer's office. "You know," he says, "I don't imagine for a minute that the things I was talking about were original to me. I was worried for awhile that you might think I was running away with Jim Luyten's and your ideas, but I was just trying them out in my own mind." I attempt to reassure him that nothing could be further from the truth: "Scientific ideas are the common property of mankind," I say valiantly. But I cannot completely hide my proprietary feelings—hide that human weakness we possess at first when we coddle our discoveries as creatures of our very own.

"It used to annoy me," he goes on, "when some younger assistant would offer suggestions about how to make a repair that I had put into his head only a few months earlier. I had to remember that I did the same thing to my elders when I was young. So, maybe you won't mind if I ask you to give me some idea of how complete you think this model really is, in your view, as an explanation of the ocean's circulation. There seemed to be a lot of things that you had to specify about the model before you could turn on that Luyten algorithm and compute the rest." I assure him that I don't mind continuing our conversation. Answering difficult questions is always easier than answering easy ones: you are not accountable for the inconsistencies. And asking simple questions is the hardest part of all.

FIRST, the model is for the long-term mean state; we do not account for how it got there. The Ekman pumping and the interfacial buoyancy flux are taken as givens, whereas a more complete model that includes the physics of the upper wind-driven Ekman layer (which we have avoided discussing in detail) might involve a lot of unknown phenomena that at present are obscure. Modeling the ocean's continuous stratification by a series of discrete layers, each of uniform density, is a gimmick, and convenient for purposes of calculation and reasoning but unrealistic. At present our model cannot be computed with more than two moving layers. Concentrating internal buoyancy gain at the interfaces is a trick that needs to be justified by deeper investigation of vertical diffusive and convective processes. All these features are in a kind of limbo. The large numerical computations done by massive computers in centers such as Princeton and Boulder encompass such phenomena, but not with all the clarity that we would like to have. In certain parts of the ocean there is, no doubt, intense horizontal mixing of properties—even of potential thickness—along the density surfaces. Many investigators put great store in the effects of mesoscale eddies, which are some 100 kilometers in diameter, and will be aghast at our leaving them out of consideration. Our only defense is: one thing at a time.

Even within the context of our model, there are quantities that we cannot help but think ought to be determined rather than taken as a given. The most conspicuous examples are the depths h and D prescribed along the eastern wall. When we know more about other parts of a more complete model, it is likely that these will be calculated, not prescribed.

Finally, we desperately need to devise some computable models of a moving, two-layer Gulf Stream system on the western side of the ocean that we can couple intelligibly with our model. You will have noted that there is a degree of freedom in our choice of h and D at points where characteristics enter the western side of the model. This will no doubt be removed when a proper model of the western boundary current is devised. A couple of years ago Jim and I published an example of a moving two-layer Gulf Stream with two control points in it, acting like two internal waterfalls to determine upstream conditions, but it is too limited and special to be coupled to our model. We hope to find a simple solution so that we can understand it, so we do not have to depend upon some misty, heavy, numerical simulation. So much remains to be done. As the old Russian proverb goes: "Little by little the water wears away the stone."

Fourteen / The Viper's Jaws

THE NARRATORS of sentimental nature documentaries are fond of reassuring us that the field mouse in the viper's jaws feels no pain—that it is insensible, in a state of shock. But I have seen old men die. The poet W. H. Yeats, in his poignant "Sailing to Byzantium," portrays the death of old men more accurately: how the soul, sick with desire and fastened to a dying animal, knows not what it is.

As Mark Twain grew older, and others in his generation began to fall away, he asserted that though others died, in his case perhaps there would be an exception. An active mind stays young, though chained to a dying body. We have been given the precious gift of life, a chance to contemplate a surpassingly beautiful universe, minds with which to cultivate some measure of wisdom, and the companionship of other creatures no less mortal than we.

If, in the midst of an often crass and strident society, we have managed to control our avarice and learned to give rather than to take, and above all to give ourselves to fellow human beings, then we may discover how, with grace, to give ourselves to death.

I WAS DISTURBED, a short while ago, to hear that the Chief has a tumor on his esophagus and is confined to a mid-Cape Cod hospital. He is the victim of a lifetime addiction to cigarettes. In his youth—long before the warnings of the U.S. Surgeon General—Uncle Sam had offered cheap, tax-free cigarettes to his soldiers and merchant seamen as a token of his gratitude. And so there, in the hospital room, the Chief sits mournfully, breathing through a tube that sticks out from his throat. The prognosis is not good. He can speak, but only in a whisper.

I suppose that it is natural for some of us to find some solace—in an overwhelmingly bewildering world—in trying to understand little things. If it is not possible to perceive a larger order, then at least we can devise a little clearing of knowledge of our own. Perhaps all of scientific knowledge has grown from our need to understand a little, if we are doomed to ignorance of the most. At any rate, so it seems to be with the Chief. Faced with desperate uncertainty, he is grasping for areas he can comprehend, and he is eager to resume our discussion of the ocean circulation.

In a faint whisper he presses his query. "I've been thinking of those great oceanic gyres that go round and round in the upper kilometer of water," he says, "and trying to picture them as flywheels. The earth itself, in its rotation around its axis, is a large flywheel that can't stop rotating because there is little to slow it down. These oceanic gyres are simply small parts of the earth that are also flywheels riding on the earth, but they would eventually lose their relative (to the earth) rotations and approach that of the earth by friction, unless their relative rotation can be maintained somehow. I appreciate the complexity of the flow patterns and density layering that is required for water in the gyres to go around without breaking the three rules. Quite clearly, it takes some intricate arrangements of structure for a liquid relative flywheel to ride on the earth without generating great accumulations of mass that overwhelm the land. I think that I do understand the diagnostic ideas that apply to local density structure, and accept that the simple deductive layered models that can be computed by the Luyten algorithm indicate how consistent patterns of density and flow can be found, given certain conditions along the surface of the ocean, on the density interfaces, and at coastlines. But I still am puzzled. Could you show me for a particular ocean, let us say the North Atlantic, just how all these ideas come together? I guess that what I have in mind is one of those diagrams that show how a gasoline engine works—you know, the intake stroke, followed by compression, then ignition, then the power stroke and exhaust. Maybe what would help would be to trace a particle of water in the ocean from the time it enters the North Atlantic circulation to the time it leaves it. As it moves along, we could examine what was making it go, and what part the oceanic flywheel plays."

So here we are again. The Chief sensed the schematic nature of our previous discussions and wants to get down to cases.

It just so happens that I am in a slightly better position to offer him an answer than I would have been a month earlier. Again, I have been enlightened a little by my computer. It is now programmed to run as an oceanographic *atlas verité*, a living atlas. For the first time, I find myself able to manipulate large quantities of oceanographic data, and to display the results in novel ways.

You remember that I mentioned my friend, the computer genius Ed Fredkin, before, and how he had outlined an idea for collecting all the oceanographic station data on a memory disk and accessing it on a color screen. We published, in 1961, a brief account of a prototype model that we had got running on a PDP-1 computer. But at that time we were unable to get it funded. Well, it seems to me that in this life you never really have to give up a good idea. Sometimes you just have to wait. Even for twenty-five years.

Now, of course, we are twenty-five years further into the computer age, and data displays are commonplace, and one does not need access to a huge computer anymore. I was determined to fabricate a working *atlas verité* that could run on an ordinary personal computer. The oceanographic data were lying more or less dead in large computer memories, and I could feed this data directly into my own little machine's memory.

I got help from Nelson Hogg, James Luyten, Ellen Levy, and George Knapp in making the data file and access programs. And in no time at all I had my own little ocean of measured data for the North Atlantic. It contains about 100,000 of the best measurements of depth, temperature, salinity, and dissolved oxygen that have been made during the past forty years. Much of it was obtained by Woods Hole oceanographers William G. Metcalf, Valentine Worthington, Frederick Fuglister, and Michael McCartney. Substantial portions of the file were made possible because of work done at sea by British oceanographers John Swallow and James Crease, and by Canadian and West German oceanographers. It is a box of gems poised for display.

Now, after a quarter of a century, I can finally study and inspect the oceanographic-station data in a flexible, interactive way. I do not need to depend on other people's editing, smoothing, contouring. I can stick right with the original data points.

I tell the Chief that tomorrow, when I come back to see him, I will bring color prints of some data displays of the North Atlantic that we can use as a basis for tying up our discussions about ocean circulation. I will try to illustrate how the dynamical rules and ideas work when applied to real ocean data instead of the simplified blocks and layers we have talked about so far.

"A little breath of salt air would do me good," the Chief replies wistfully.

When I get back to my lab I call up on the screen a perspective picture of the North Atlantic, looking obliquely down on it from a position high over southern Africa. I choose some vertical sections along which ships had made systematic, closely spaced observations. The geographical location of the sections that I choose is shown in Figure 14.1. Then I paint the data points onto the vertical sections. They look like a set of intersecting bead curtains, each bead a real data point. I cannot put all the historical sections in the display because that would produce an undecipherable cloud of points.

I paint the first display so that each bead is assigned a color according to the observed density. After all the colors are assigned I can see the density stratification standing out by the bands of color. In this way, I can portray in perspective how density of water is distributed geographically and in depth throughout the North Atlantic. There are many subtle details that are missed by the coarse grain of this particular display. I then repaint the beads with colors according to observed temperature, then salinity, and finally to observed dissolved oxygen. These are the four displays that I will take to the Chief. You might wonder why I do not make up similar perspective pictures of the other observables such as the chemical concentrations of phosphate and nitrate and the more subtle radioactive isotopes and decay

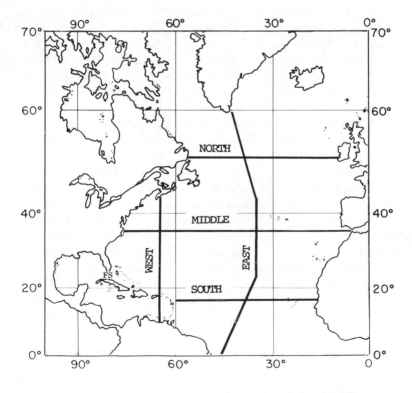

FIGURE 14.1. The location of the hydrographic sections depicted in Figures 14.3–14.5

products such as tritium and helium-3. The answer is that the data on these quantities are pretty sparse. The bead curtains would have few beads indeed, and the color would be spotty. We would have to devise a different type of data display for them.

The displays of original data on the computer screen are stunning in their bright colors, but I would have to sketch something on paper for the Chief's sake. By comparison, my hand-drawn sketches will be pallid indeed, and smoothed of much interesting detail. I choose five vertical sections of data in the North Atlantic for illustrative purposes. Their locations are shown in Figure 14.1: two are more or less north-south sections made by McCartney fairly recently, and three are east-west sections made in the 1950s by Worthington and Metcalf. They go from coast to coast. We will look at the properties of the upper 1000 meters of these sections. The location of the Florida Straits, from which the Gulf Stream issues, is shown in Figure 14.1 by the letters FS. I would like to use a lot more of the sections that my computer is projecting on the screen, but I cannot crowd them all together on one piece of paper. The sections I choose are at about 64°W and 36°W longitude, and 53°N, 36°N, and 16°N latitude. Now, we imagine that we are at some point high over southern Africa and looking northwestward over the North Atlantic. We see the upper 1000 meters of the five sections in perspective. They look like a grid of intersecting walls or fences.

From the computer display I trace off contours of temperature, salinity, density, and dissolved oxygen, and then color them in, guided by the computer screen. Of course, I really do not need a computer for this particularly easy job—it could all be done by hand or partly by copying from published atlases—but it would also take at least ten times longer. And suppose that in the course of tracing I discover that I want to rotate the perspective a little: days of work by hand, almost instantaneous on the computer. Then I color in the spaces between the contours with colored pencils. This is a delightful kindergarten type of exercise—one I think you will not despise, because it is so instructive.

The results of this activity are shown in Figures 14.2 to 14.5. They show the static distribution of properties in the main part of the circulation of the North Atlantic. For the sake of convenience, let us refer to the sections in the following manner: the 64°W longitude section we will call the west section; the 36°W section the east section; the 53°N section the north section; the 36°N section the middle section; and the 16°N section the south section.

Let us now turn to Figure 14.2 and study its temperature structure.

116

On the north section the temperature is mostly less than 5°C except for a fairly deep region on the eastern end of the section where the water is warmer—in fact, it is higher than 10°C. We can see in Figure 14.4 that this is a region of lower density, too. This means that a component of the HPG is directed westward along this section in the lower density water, and that this is most probably balanced by the CF as-

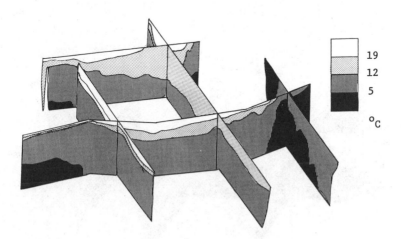

FIGURE 14.2. Temperature in degrees Centigrade on the five sections shown in Figure 14.1. The top 1000 meters only. View is from above Africa looking toward the northwest. Contours smoothed to convey basinwide view of distribution.

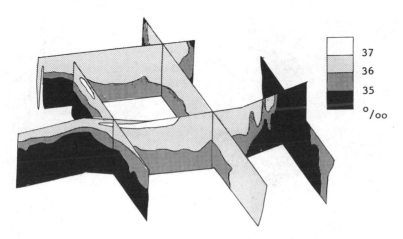

FIGURE 14.3. Salinity in parts per thousand in the top 1000 meters.

sociated with a northward component of GV in this top layer. This is the branch of the circulation in the North Atlantic that brings heat up along the northern European coastline and releases it to the air, thus ameliorating the climate there. According to Figure 14.3 the salinity of this water is higher than the water under and to the west of it, and as we will see, this implies a southern origin for this water. The oxygen is high, which suggests active interaction with the atmosphere to

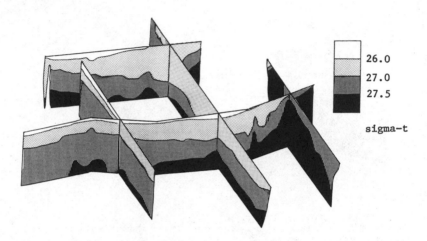

FIGURE 14.4. Sigma-t in the top 1000 meters.

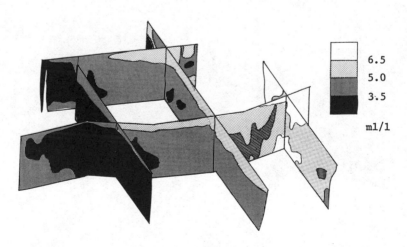

FIGURE 14.5. Dissolved oxygen in ml/l in the top 1000 meters.

depths of several hundred meters. The oxygen is even higher further to the east on this northern section because the temperature is lower: the colder the water, the more oxygen can be absorbed from the atmosphere (Figure 14.5). The western portion of the north section appears to be nearly homogeneous off Labrador in all properties except in salinity. A shallow layer of very low salinity, known as the Labrador current, hugs the coast, flowing southward across the section and carrying low salinity because it accumulates melt water as it passes along the frozen arctic coasts.

Now we follow the east section from its northern end at Greenland down southward across the northern section. Immediately after the crossing, we encounter higher temperature, higher salinity, and lower-density, lower-oxygen water that extends to depths of about 800 meters. From our ideas about HPGs and GVs we can assume that this structure is evidence of an eastward-directed current in the upper 800 meters. If we use our computer *atlas verité* to explore the region east of the east section and south of the north section, we will find that this current is more or less continuous with that flowing north across the northern section. The temperatures and salinities on the upstream section (east section) are higher than on the northern section, which suggests that considerable cooling by heat loss and freshening by rainfall occur between the two sections across this northern Atlantic current. The curved arrow connecting the two portions of these sections in Figure 14.6 shows the location of this North Atlantic current.

We proceed southward along the east section. As we do so we encounter a warm pool of water exceeding 15°C. In fact, the 10°C isotherm reaches down to about 700 meters depth as we cross the middle section. This is part of the "warm pool" in the center of the subtropical gyre of the North Atlantic. Here the salinity is high and the oxygen is lower than in the northern section. The density stratification is high, which effectively isolates the various layers from one another. The downward slope of the density surface toward the south along this portion of the east section is indicative of a strong eastward current in the upper layers, part of which turns toward the south along the European coast and crosses the middle section east of the east section. This is also indicated by an arrow in Figure 14.6. The arrow curves around clockwise and forms part of the great subtropical gyre. It then continues to turn clockwise and crosses the east section again south of the middle section, but this time it goes westward. When we examine the density structure on Figure 14.4 along the east section between the middle section and the south section, we discover no pronounced slope in the deepest density surface (at a sigma-t of 27.5) and therefore no pronounced current at 1000 meters. Sigma-t is a shorthand unit for density that oceanographers like to use. It is given by the formula sigma-t = (density − 1) × 1000. However, the density surface 27.0 does have a slope up toward the south along the east section, and this is indicative of the westward-flowing current there. Evidently this current is not much deeper than 700 or 800 meters. It goes by the name of North Atlantic Equatorial Current, but it is quite a bit north of the equator. Its shallowness suggests that the lower layers at this location are in the shadow zone. The oxygen at 700 meters on the east section, where it crosses the south section, is quite low. The conventional interpretation of low oxygens is that the water has been subsurface for so long that oxidation of detritus falling from the surface has used up the oxygen measurably. Slow-moving or stagnant water is indicative of the presence of a shadow zone. We encounter even

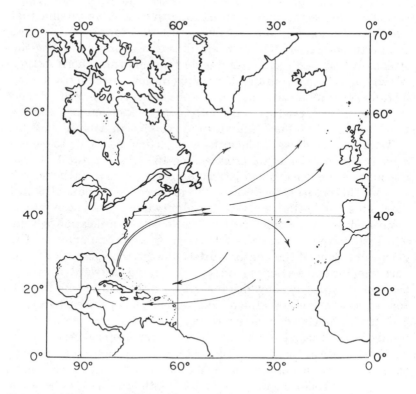

FIGURE 14.6. Average direction of currents across the hydrographic sections as inferred from slopes of the density surfaces in the upper 1000 meters.

lower oxygens at mid-depth of our 1000 meter-deep sections on the eastern segment of the southern section, farther into the shadow zone.

THE CHIEF interrupts: "I'm not sure that I like your drawing those big arrows in Figure 14.6 to show the currents across various parts of the gyres. Aren't you in danger of implying that the direction of the velocity is the same at all depths across the sections? That would be a retreat to the old, now discredited, law of parallel solenoids."

I feel a little chagrined. He has me again, all right. I falter a bit and tell him that if we just take those big arrows to indicate the vertically summed flow, we will be O.K. The direction will be the average direction. We can get back to the spiral part later. The answer is not really good because there ought to be a better way to draw the diagram, but I do not want to lose the ground I already covered.

WE NOW CROSS the southern section and find ourselves in a region where the temperature is colder at depth again, and the salinity is lower, too. Part of this low salinity is due to the water at intermediate depth that comes up from the southern South Atlantic at intermediate depth. The upper layers are still going westward in the southern flank of the North Equatorial Current.

Now let us follow this westward current to the west until we encounter the west section. Part of the current goes through the Caribbean and some continues west to the north of the Caribbean in the open Atlantic. The part that flows through the Caribbean then enters the Florida Straits, which are too narrow and shallow to show well on this perspective diagram. The part of the current north off the Caribbean joins the Florida Straits current to form the Gulf Stream. You will notice on the west section that the pool of warm, saline, low-density water is very deep at mid-latitude, where the middle section crosses the west section. Inshore of this point of intersection, the middle section reaches to the coast and crosses the Gulf Stream. This is the first evidence we have seen on our sections of the steep slopes of temperature, salinity, and density surfaces across this stream. The 15°C isotherm cuts the sea surface, keeping the warm central pool away from the coast. Because of the large slopes, the speed of this Gulf Stream current reaches 2 meters per second, compared to speeds of less than 0.02 meters per second in other portions of the subtropical gyre. As the Gulf Stream follows a path some distance offshore, we observe it cutting through the west section in an eastward direction, at a latitude a little north of the middle section. It is still going strong, and we have available quite a few nice sections of it as it hurries along here, but they would just clutter up our diagram if we tried to include them in this presentation of our very sparse selection of data.

We have now completed a tour of the sections. Now we can get down to that business of tracing a little cloud of water particles through the ocean system.

I had prepared a series of similar diagrams to allow us to trace the progress of particles of water as I imagine it to be from the time the particle enters the ocean circulation to the time it leaves it. These diagrams are shown in Figures 14.7 through 14.12. As we follow the trajectory of our particles on these figures, we will need to refer to Figures 14.2 to 14.5 to identify what layers they are passing through.

Let us begin with Figure 14.7. It shows the direction of the drift of the wind-mixed layer that is forced by the prevailing winds. In the

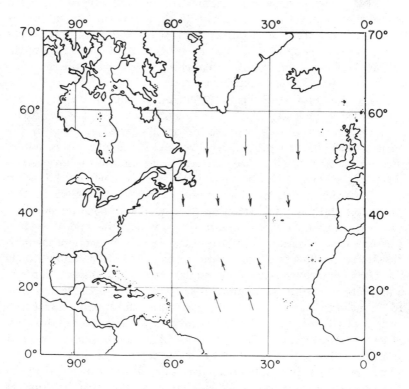

FIGURE 14.7. General direction and amount of flow in the wind-driven Ekman layer as inferred from prevailing directions of the winds. The place where the arrows converge is the major region of downward Ekman pumping.

northern portion of the map, the winds are blowing toward the east and the Ekman drift is southward. In the southern portion of the map, the winds are the Trades, blowing more or less toward the southwest, and they drive the Ekman drift of the mixed layer toward the northwest. These two opposing drifts meet at mid-latitudes. The water of the mixed layer is forced downward by the convergence of the surface drift; this constitutes the Ekman pumping all over the subtropical gyre. We now consider a little cloud of particles at point 1 in Figure 14.8. Our cloud of particles is in the wind-mixed Ekman layer, is thoroughly heated up, and has a salinity that is pretty high compared to the rest of the ocean, caused by the excessive evaporation at this latitude. The surface, at mid-latitude, is an important source of high salinity. The cloud is now forced down by the Ekman pumping into the density layer immediately underneath the mixed layer. It is forced to move southward by the general direction of the flow beneath it at point 1, and forced to sink by subduction under lighter water at the surface as it moves southward. It gets down to about 150 meters and is visible on the east section as a maximum in subsurface salinity close to the top surface. Then it enters the Gulf Stream and is carried rapidly north (point 2) and then east into a cooler climate. It also outcrops as it leaves the Gulf Stream and loses some of its heat at point 3. Getting denser, it flows across the density surface underneath it and enters an even denser layer, and gets a little Ekman hat as well. It now subducts in this layer and turns southward again to recirculate through the subtropical gyre as shown in Figure 14.9. Its path is not quite the same because of the beta-spiral, and so it moves a little to the right of the water above it. It does not go quite so far south the second time around. It now passes underneath its first trajectory at point 4, in a deeper density layer. It then re-enters the Gulf Stream and is carried even farther north this time before it outcrops, loses heat again, and sinks into a denser subducting layer at point 5. It recirculates like this several times, each time getting into a denser, deeper layer and outcropping and subducting at increasing latitudes (Figures 14.10 and 14.11). Each time it outcrops, it picks up some fresh water from precipitation at high latitudes and gets well ventilated with oxygen from the air; it also moves more slowly through the gyre. This goes on until the remnant either flows into the Norwegian Sea far to the north and eventually gets converted to bottom water (point 8), or until it wanders into the Labrador Sea where it gets cooled and freshened even more (point 9). Finally, some convects to levels below 1000 meters and flows along the western boundary of the North Atlantic down past Cape Hatteras and finally into the South Atlantic.

"I THINK that I now grasp your view of the circulation," says the Chief. "Water enters the gyre in two ways: either by direct forcing downward by addition of Ekman hats through the agency of the wind, or by winter-time convection in more northern portions of the gyre, where the buoyancy loss associated with cooling causes water to cross the mean position of the equal density interfaces. The former is confined to a rather shallow surface layer in the subtropics, whereas the latter occurs at greater depths—up to 600 meters—in the subpolar regions. The heat that is lost from the ocean here originally comes in with the Ekman-hat water in the subtropics. The heat actually enters

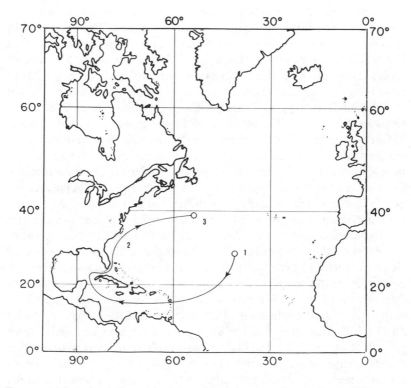

FIGURE 14.8. A cloud of particles injected into the upper density layer of the ocean by Ekman pumping at point 1 (where this particular density layer outcrops in winter) sinks to a depth of about 150 meters and moves through the Caribbean into the upper layers of the Gulf Stream (point 2); it then cools and sinks into a layer of slightly greater density at point 3, at a location in the wintertime outcropping of this new density.

the ocean in the mixed layer of the subtropics before it gets pushed downward in the form of Ekman hats.

"Once the water has entered the circulation from the Ekman layer, it is confined more or less to density layers, moves into the Gulf Stream where it flows rapidly northward, and tends to break the surface again. These density layers aren't flat—in fact, they tend to slope up toward the north in subpolar regions, and they outcrop there. As a particle encounters a cooling region, it sinks down to a deeper density layer at the outcrop, and is joined by some more particles that come down as new, somewhat denser Ekman hats. The flow southward and

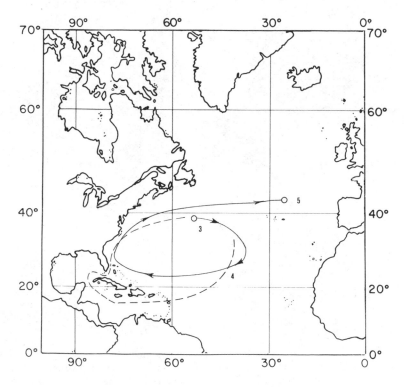

FIGURE 14.9. From point 3 the cloud of particles, now augmented by some more Ekman pumping, flows southward, again in a deeper layer—about 300 meters—and at point 4 crosses underneath its previous path, this time moving somewhat to the right of its previous direction. This is the upper part of the beta-spiral. It then rejoins the Gulf Stream deeper down and emerges at point 5 in a region of wintertime outcrop.

westward in this new density layer includes the recirculated original particle and the new ones that were added while cooling. This flow around the gyre does not penetrate as far south as that in the shallower layer previously traversed, and leaves an unventilated region in the tropics that you call the shadow zone. Eventually, the cloud of particles enters the Gulf Stream again, this time deeper down in it, and when it reaches the Grand Banks it cools some more and sinks to a deeper density layer. Again it is joined by some new Ekman hat particles during winter-time convection. The number of particles in the

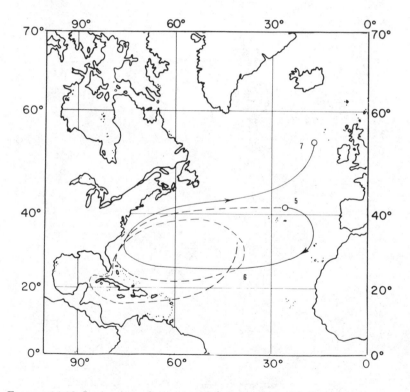

FIGURE 14.10. Losing more buoyancy while in contact with the wintertime convective process, the cloud of particles sinks to an even denser layer at point 5, moves southward again—but not so far south as before—as it tries to avoid the shadow zone in this layer, and then crosses underneath previous paths at an increasingly right-turning direction at point 6. It then enters the Gulf Stream at a deeper level, and reemerges even farther north at point 7, where it again becomes denser. These circuits are transporting heat to the north of the ocean, where it is lost to the atmosphere for the sake of the European climate.

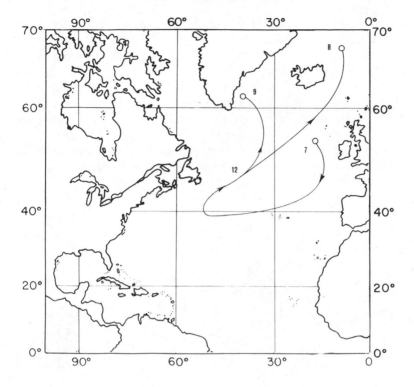

FIGURE 14.11. The cycle is repeated, the gyre is farther north, and the cloud splits at point 12. Some of the cloud of particles passes into the Greenland Sea and sinks to the bottom at point 8, and some sinks at point 9 without crossing the ridge between Iceland and Scotland.

cloud is increasing, but the thickness of the layer is also increasing, and the speed with which the cloud is moving decreases. I can think of the density layers as flywheels tipped up toward the surface in the north, where at every revolution (slower in the deep layers) some new particles are added to the cloud. When the cloud gets far enough north, the Ekman layer actually extracts particles from the cloud when it surfaces at the outcrop and shoves them southward in the top wind-mixed layer to resupply the Ekman hatting in the subtropics. It has some difficulty doing so because the wind-mixed layer is continuously short-circuited by its contact with the layers below. Just looking at this picture, with the wind-forced Ekman hats and the convective transfer of particles from one layer to another at outcrop, gives me

127

some feeling of how both the mechanical driving of the wind's stress on the sea surface and the buoyancy transfer from low to high latitudes give a nudge each time around to density layers when they outcrop and keep the flywheels going.

"Somehow it's comforting to me, a man who has spent his life at sea as an engineer, finally to get some kind of idea of how the ocean itself works. I suppose that you scientists will change your minds about these ideas in the years to come, but they'll have to do for me."

"Still," he goes on, and his voice is weak and tired, "I can't help saying that what you've been showing me today is a weird mixture of fact

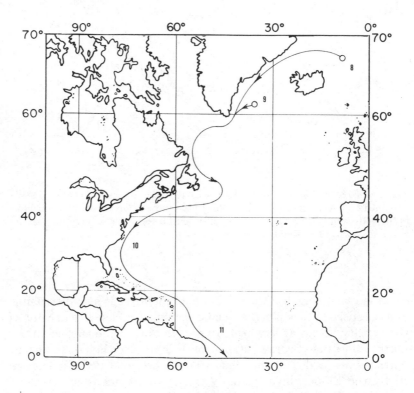

FIGURE 14.12. The very cold water from point 8 fills the Greenland Sea until it brims over the ridge between Iceland and Greenland and forms a narrow deep current hugging the eastern side of Greenland; more joins this flow from point 9. These deep waters flow in a deep western boundary current along the coast of North America, inshore of the position of the Gulf Stream. At point 10 the cold water crosses under the Gulf Stream and eventually ends up in the South Atlantic (point 11).

and fancy. That business about the computer atlas and the data it displays is fact, all right, but that narrative about how a cloud of particles circulates around in the ocean seems like stretching things. The beta-spiral ideas in the middle of the gyre may be O.K., but do you know what really happens at the base of the wind-mixed layer, and how the water presents itself potential-thicknesswise to the water below? Are you confident about the physics of the winter-time convection and the subduction processes? Isn't your talk about the subpolar gyre perhaps oversimplified? What bag of tools and tricks are you going to use at sea to find out more about these rather poorly defined processes? And certainly there must be an important role somewhere in the machine for those eddies that you studied fifteen years ago in the project that was called MODE and the one with the Russians called POLYMODE. I hope you find a way to sharpen up the knowledge in those areas."

What can I reply? When faced with the unknown, we speculate, we seek an entry point into an unknown territory, and at times we feel despair. The Chief is right. There is so much to know and learn, and we must grope around for ways to find it out. Even the problems have to be defined.

There are, in our society, a group of solid, sensible men who make it go. They are the petty officers in the Navy, the airline pilots, the master printers, the devoted surgeons, the best farmers, the builders, the artisans in the machine shops. They are that segment of society that Nevil Shute celebrated in his novels. You won't meet them advising the President. They don't go to academic cocktail parties. They seem so competent within the confines of their fields that we tend to think they lack imagination. They observe the passing scene and they don't talk much. Perhaps, really, there is little, in the end, to say.

The Chief's fate is sealed and he knows it. Within a few weeks his tarred lungs are clogged with pneumonia. I see him one last time, and he is very weak. He manages a smile and whispers, "Thanks for giving me something pleasant to think about."

His ashes are scattered in the soft blue water off Bermuda. As they sink and dissolve they enter the great ecological cycle of the ocean—first as soluble phosphates, as nitrates that fertilize the phytoplankton. Then the zooplankton eat the phytoplankton, larvae eat the zooplankton, and so up the zoological food chain to perhaps a bonita, caught on the hook trailed behind an oceanographic vessel by an engineer who likes to fish. I will remember my Chief when I pass that spot off Bermuda and pour an appropriate measure of Methuselam rum into the sea.

Appendix / Computer Programs for the Reader

IT OCCURRED to Jim Luyten and me that a serious reader might like to have some programs to run on a personal computer with a color card and color monitor. We offer here nine little programs in Micro-Soft BASIC that are suitable for an IBM/PC, or a similar machine such as the NEC-APC III.

These programs are meant to be fun. But they are also useful in learning about the Coriolis force and the idea of a beta-spiral cruise; the computer carries out automatically the kind of diagnostic constructions that the Chief had to do by hand. In addition, they will run versions of the deductive models computed by the method of characteristics. They are supplemental to the text and to my discourse with the Chief. Even though they may be fun, they are not "games." The first three are more or less educational. PROG4 to PROG6 are serious diagnostic tools that you could apply to real ocean data. PROG7 and PROG8 are research tools that compute theoretical models that could be cumbersome and difficult to calculate in any other way. Some of our analytically minded colleagues may prefer to ignore them. But why walk to San Francisco when you can fly, and get a good view of the scenery on the way as well?

To use these programs, you must first boot up your computer with MS-DOS and then load MicroSoft GW-BASIC.

The programs are mostly self-explanatory, and of course the text outlines the general ideas. However, some comments and directions are in order. We have arranged the programs so that each is completely independent of the others—you do not have to copy them all to make one of them work. But because you may like the idea of building up a complete set, we have also included an introductory Table of Contents (PROG0) that will lead you into them automatically. It probably is not a good idea to merge all the programs together in one single large program because then you might exceed the internal RAM of your computer. Therefore, although the numbering does not overlap, and the programs could be merged, we have arranged the Table of Contents so that it loads and runs each program according to its label—for example, PROG1, PROG2, etc.

In the programs, we refer to the colors by number. This is because

131

the palettes of different computer models are not always the same; moreover, you may want to change your palette sometime. If you have an NEC-APC III with seven colors, you may want to enrich your displays by changing whatever MOD 3 instructions you find to MOD 8. Then, of course, our color numbers will not apply anymore. If you are not sure what the color numbers mean in terms of your own machine, choose PROG9: it will display the color numbers with the colors that your own machine is using (unless you have altered the programs).

PROG1

Particle rotating around the earth's axis and gravitating toward the earth's center. Refers to Chapter 2 and Figure 2.1. This program shows, in absolute space, a pendulum bob (color 8) suspended from your ship on the oblate spheroidal rotating earth (color 1) and being accelerated toward the axis of the earth's rotation. The machine will ask you to choose a value for the spin of the earth. If you choose spin = 0, then you will get a resting spherical earth; but if you choose a number up to spin = 1, then you will get an earth that is oblate. At spin = 1 the earth flies apart—this is the limiting spin for a self-gravitating sphere. So choose some value between 0 and 1. The large ellipse is supposed to represent sea level on a meridional plane cutting the earth, as in Figure 2.1. The little color-3 line shows how a plumb bob (the color-8 ball) suspended from your ship would hang down for various spins. You will notice that when it is referred to, the perpendicular to sea level always appears to point straight downward. However, as the spin gets bigger, it really points farther and farther away from the center of the earth. You can think of it as being thrown outward from the axis of the spinning earth by a "centrifugal force," but it is better to think that this really means a centripetal acceleration toward the axis; this is represented by the color-8 vector acting on the color-8 plumb bob. The tension in the string acting on the bob is represented by the color-3 vector. The gravitational force on the bob is indicated by the color-2 vector and points approximately toward the center of the earth. This is only an approximation because a spheroid, unlike a sphere, does not attract exactly toward its center. The gravitational force and the tension in the string do not balance. They are the only forces acting on the bob (because the "centrifugal force" is fictitious). So the force shown by the color-8 vector is the resultant, and this is what gives the bob of unit mass its axipetal acceleration, which, in the case of unit mass, can be represented by the same color-3 vector. Try some different values of the spin, and get a feeling for how the direction of the local vertical depends on the spin. Think about what this means for the definition of local latitude.

PROG2

Particle sliding on a frictionless ellipsoidal earth. This program illustrates the notion of a particle subject only to Coriolis force and supplements Chapter 3. It shows you two views of the earth. On the left you see it from a distant point that is fixed in absolute space—in the Newtonian sense. On the right you see the earth from a point that is moving with the rotation of the earth, so that it is always directly over the same geographical point. A geosynchronous satellite won't do unless you want to sit over the equator. The circles of latitude of the earth are color 1.

The program first asks us to choose a latitude. Suppose that we choose 45°N; enter 45 and type RETURN. Then it wants to know what relative velocity to give to the sliding particle. The scale is set so that ten units correspond to the earth's absolute velocity at that point on the earth. The eastward relative velocity is u and the northward is v, so you have to enter the two components. Now, after you hit RETURN, the display begins. The starting point of a sliding particle is marked with color 2. As you will see, this point, fixed as it is geographically, rotates with the earth as seen in the absolute view, but it seems to remain fixed in the relative view on the right. It tends to be obscured by the particle's trajectory on the right-hand view. The sliding particle has color 3. It moves in both reference frames.

If you begin with lat = 45 and $u,v = 0,0$, you will verify the fact that the particle stays at a fixed geographical point if it rotates with the earth. The slope of the ellipsoid up toward the equator tends to make the particle slide toward the pole, but this force is precisely enough to accelerate it around the axis, and the particle never does move toward the pole, but only circles it. This is the equilibrium shown in PROG1. (Incidentally, we do not actually show the ellipticity in this figure because the earth is too small to see on this scale and resolution.)

On the other hand, if we enter lat = 45 and $u,v = 2,0$, the particle will begin with an eastward velocity relative to the earth, and the down-slope force that the equatorial bulge exerts on the particle will not be sufficient to accelerate it around the earth's axis; therefore, the particle begins to slide out toward the equator to a higher place on the bulge—a lower latitude. When I tried to explain this to the Chief, he said, "Sure, just like the balls on a steam-engine governor."

You can see this happening in absolute space. The particle sepa-

133

rates from the geographically fixed point on the sphere, where it would have remained if at rest with respect to the earth, and moves first ahead—toward the east, faster than the reference point—and then turns southward as it mounts the equatorial bulge. In doing so, however, it gets farther away from the axis; trying to conserve its angular momentum, its eastward velocity slows down, until it finds its eastward velocity relative to the earth has fallen off to zero and moves southward in the relative space. Then it overshoots this latitude and gets so far from the axis that it now has a relative velocity that is westward—less than that of the earth in absolute space. By the time this happens, the relative southward velocity has dropped to zero; in other words, it has mounted up the equatorial bulge as far as it can go, and now it has to slide down northward again. As it does so, it regains its eastward relative velocity and slides back to its starting latitude again, where it again finds itself with an excess of eastward velocity. The unfortunate thing is thus doomed to skitter around the globe forever. When the amplitude of the skitter is reduced, say, $u,v = 0.5,0$, and the latitude is high, say, lat = 70, then the particle tends to pass through its starting position, with a period of the earth's rotation in absolute space divided by twice the sine of the latitude. The trajectory is then nearly periodic and approximately circular. These are the so-called *inertial circles*.

Seen from the point of view of an observer on the relative frame— that is, a bystander positioned on the earth on a tower where he or she can plot the velocity of the particle relative to him or her, and its position—it will seem that the particle's trajectory can be explained by a simple horizontal force, acting to the right of the velocity of the particle and proportional to the product of the relative velocity and the Coriolis parameter, which is itself the product of twice the earth's angular velocity and the sine of the latitude. This is the so-called Coriolis force.

The trajectories need not be circles. If we choose lat = 30 and $u,v = 2,0$, the trajectory will cover quite a range of latitude, and therefore of the Coriolis parameter. The trajectory will not close, but it will gyrate in a clockwise circle drifting toward the west. If you enter even larger values of u,v, say, 4,0, then the particle will mount so high on the equatorial bulge that it will run over into the southern hemisphere, where it will speed up again toward the east, and then get thrown over the equator again into the northern hemisphere. It therefore executes a wavy trajectory skipping over the equator.

It is a good idea to study variations on these particular examples, looking at the movement of the particle in both absolute and relative frames simultaneously, and getting familiar with the apparent differences.

This setup enables one to get a good intuitive feel for Coriolis forces. As you will see, they are the only force that acts upon the particle, so far as that fellow standing on the tower is concerned. When we consider the motion of ocean water and of the atmosphere, there will be other forces, particularly horizontal pressure gradients that can balance the Coriolis forces.

PROG3

Beta-cruise, discussed in Chapter 4. This program permits you to take a beta-spiral cruise from Las Palmas to Cadiz. When you press c to start it running, a map (10-degree grid color 1 and land masses color 2) of the North Atlantic will appear, with a 10-degree grid of latitude and longitude. The two blinking lighthouses (color 3) are at Las Palmas and Cadiz, your scheduled ports of departure and arrival. You have twenty days to do your whole survey and get into port again. You will also see the outline (color 1) of the triangular area that you wish to survey, and a large black blank portion of the screen on the top labeled "HYDROGRAPHIC SECTION." A marker (color 1) moves from right to left across this region during the period of the cruise. It is here that one plots the observed depths of five density surfaces, indicated by different colored lines.

You will observe that there is a day count at the bottom right of the chart showing time still left. A color-3 line starts to move southwestward from Las Palmas. This is your slowly moving ship's track. At any moment you can change its heading by pressing H and entering a new compass heading (000 is north, 090 east, 180 south, and 270 west). Press RETURN, and the cruise resumes. You may want to alter the heading to steer your ship toward the southeastern corner of the triangle.

At any moment you may encounter a storm ("THREE DAY N.E. GALE") that sets your ship to the southwest, upsets your plans, and uses up three days of precious ship time. It comes at a different time every time you use the program. It is one of the vicissitudes of planning an oceanographic cruise. That is life at sea.

When you arrive at a point where you want to determine the depth of the five density surfaces, you take a CTD station by pressing s. You can mark this point on the section display and chart with a color-1 line and small square, respectively, by pressing M. Taking a station costs

135

you two hours of ship time. The marker is free. Try to steer the ship all around the triangle, and see if you can get some stations inside it too.

While you are busy at your survey, the Captain will be plotting up the distance from Cadiz, and when the time comes for you to depart the area of survey so that you will arrive in Cadiz on time, he will order the ship to steer for port. You cannot make anymore measurements. Another fact of life.

In a real cruise, time can be wasted repairing faulty gear and overheated winches, or taking a sick crew member to the nearby Azores just north of the triangle. You have been mercifully spared these multiple contingencies.

Play around with this cruise plan, and see how Dave Behringer and Henry Stommel spent their five cruises out there.

As you go around the triangle, the observed slopes of each density surface will vary from one another. That makes the whole system lock together in a "Chinese Puzzle" under our Rules 1, 2, and 3. Better take another look now at Chapters 5 and 6.

PROG4

Diagnostic calculations. Shows three successive states of the display for a computer realization of the diagnostic constructions described in Chapter 7. When the display starts, the block of water appears on the right side of the screen. The display has default values for h (color 2), D (color 3), and B, the bottom depth (color 1). These interfaces are initially level.

The quantities $g1$ and $g2$ for which the machine asks you are the density differences across interfaces that you can choose (in parts per thousand). So enter something like 1,1. The machine then wants to know what values of h, D, and B should be assigned to the three nearest corners of the two interfaces and the bottom. If we want to model something like the North Atlantic off Ireland, we can do the following: for $h0,h1,h2$, use the values 0,200,200; for $D0,D1,D2$, use 400,400,600; and for $B0,B1,B2$, use 2400,3600,3600. Then the computer wants to know if these have been entered correctly and your answer might be yes (y). It now wants to know the latitude of the center of the block, which we can enter as 55, and your guess of the speed, v, and direction, T, of the current in the third layer down (between the color-1 bottom and the color-3 interface). You already learned the main ideas in Chapter 7, but you will find here that it is best to choose a fixed v, say $v = 1$, to begin with, and then to vary T. Try various values of T until the interfacial flux across the color-3 interface, w_2, vanishes. Something like $T = 278.5$ should work fine, but try other values first, and try to find the answer yourself. Then you are still left with w_l and w_e to adjust. Suppose you know that the Ekman pumping at this block is negligible. You should set $w_e = 0$ by trying various values of v. Try a few (answer is near -0.4), leaving T always the same. Now you can fix nothing else and you are stuck with a fixed value of w_l. This interfacial flux, which turns out to be downward (negative) in this particular example of the subpolar gyre, as shown on the display, means that upper-layer warm water is being converted to cooler second-layer water at a known rate. This implies a certain rate of heat loss from the ocean into the atmosphere (the famous warming of Europe by the North Atlantic Ocean). We have computed it from knowledge of Rules 1, 2, and 3, the observed oceanic density field, and a rough estimate of the Ekman pumping due to winds. This is our little climatological triumph. The computer keeps track of the GVs in each layer as well.

Try some other oceanic regions that have good slopes of two-density surfaces.

PROG5

A diagnostic model with four interfaces (or three interfaces and a bottom). This tool enables you to set the slopes of four surfaces, and the density contrasts $g1, g2, g3$ across the top three surfaces, for a beta-triangle such as described in Chapter 7. After entering the latitude at the middle of the block, you are invited to choose a speed v and a direction T for the flow in the bottom layer. First, find the value of T for which the interfacial flux across interface 3 goes to zero. Then find the value of v for which the flux across the next interface up vanishes. Then, unless you are lucky and the data you have chosen for the slopes and density differences are perfectly tuned to the three rules, you will be in trouble with the flux across interface 1. If you want to play the game in the spirit of the beta-spiral calculations, you will try to find a v and T for which the sum of the squares of the three interfacial fluxes is at a minimum.

If you are copying this program from the listing, you will notice that there are long portions that are duplicated in the previous program. It would be more convenient for you to write over the previous program than to type in every line anew.

PROG6

This program is much the same as PROG5. Again, if you are copying it from the listing, you will save effort by writing it over a listing of the previous program. In this program we have already chosen the

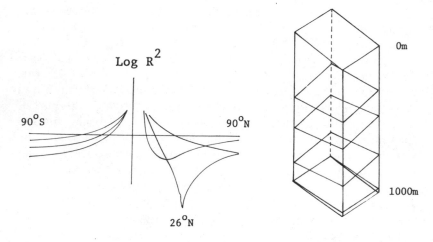

FIGURE A.1. A trial calculation using data from a beta-spiral survey showing the absolute minimum of the sum of the square of the residuals (weighted by latitude squared).

slopes of the four interfaces from a graphical fit to the slopes of one of the beta-spiral cruises, as determined by plane fits to all the data on that cruise. We are going to see how well tuned to latitude the beta-spiral data is—as described in Chapter 7. (Figure A.1) The program has already been set so that the direction T of flow in the lowest layer satisfies the condition of no interfacial flux through either of its bounding interfaces. The program then tries various values of v and latitude by itself, each time plotting the logarithm of the sum of the squared interfacial fluxes across the other two interfaces. You will be surprised to see that in this case the lowest value of the quantity log R^2 occurs for a slightly wrong latitude. This seems to suggest that there really might be some flux across the interfaces; a detailed study has not been made, but this may be evidence of cooling by downward flux at intermediate depths in the thermocline of the subtropical gyre. This is a subject for careful future study. We bring up the matter here only to show how study of both diagnostic and deductive models may lead to detection of hitherto unexpected processes deep within the ocean.

PROG7

The model using the method of characteristics. Refer to Chapter 10 on the Luyten algorithm. This is one of several working models of the subtropical oceanic circulation that can be used to study how fixing the fields of w_e and w_l in a rectangular shaped ocean determines the depths of the interfaces and the currents in a model ocean. The theoretical basis of the method is described roughly in Chapter 10.

The first display on the screen asks whether you want to use the default values of the Ekman flux w_e and interfacial flux w_s—the functional form in x (longitude) and y (latitude) is already fixed as trigonometrical; the depth D_0 of the lower interface on the eastern wall; and the latitude of the outcropping of the top interface y_s; or whether you will accept the default values of these parameters. We suggest that you choose the defaults the first time around. So strike c for continue. Now three panels appear. The left-hand panel shows the shape of w_e with latitude. It is a sine function, with strongest negative value at mid-latitude 30°N. The latitude of the top side of the box is 40°N, that of the bottom side 20°N. The little square ocean is 20 degrees wide in longitude.

The two other panels show the form of the heating-cooling function w_s acting on the top interface. It is a sine function too, but of half the wavelength of the w_e. It is positive in the southern half of the ocean, which means it is heating, and transfers water from the cool layer to the warm layer across the top interface. Maximum heating is at latitude 25°N. It is negative in the upper half of the box, but of course it cannot act on the interface north of where it happens to outcrop. The color-2 latitude line is the position of the "subduction" latitude y_s that you have chosen. When you have confirmed that everything has been entered into the program the way you want it to be, strike c again, and another set of three panels will appear.

The color on the left panel indicates the density of the water being forced downward by the Ekman pumping (this is set by choosing y_s), with color 1 for cold, color 2 for warm. The right panel shows how heating at the upper interface is distributed by w_s, with color 1 for cooling and color 3 for heating. You will now observe a field of color contours building up in the center panel. The boundaries of the colors are the contours of the Sverdrup transport function $h^2 + D^2$. You will note that this is entirely governed by the w_e forcing and not at all by w_s. Strike c again, and once again three panels appear, but now we are in the mode to calculate the characteristics and the final fields of h and D.

We observe curves shooting out from the western edge of each of these boxes, then curving around southward, and then westward again. These are the characteristic curves emanating from the "Gulf Stream." At first they follow the path of the Sverdrup contours we just saw; but after they cross into the region south of the outcrop, they veer

sharply westward, coasting up the Sverdrup contours. The characteristic curves are the same shape in each panel, but they have different
colors. What do the colors denote? On the left panel the colors denote
the depth of the upper interface h; on the right panel they denote
depth of the lower interface D; on the middle panel they denote the
sum of the depths, $h + D$. This last panel is useful because the contours of $h + D$ indicate the direction of GV in the top layer, whereas
contours of D indicate direction of GV in the lower layer.

After the characteristic curves from the western boundary have
filled up as much of the ocean as they can, new curves come from the
eastern wall and fill up the rest. The color boundaries are contours of
$h, h + D$, and D.

You must refer to the text to discover the physical meaning of all
these pretty, colored pictures.

PROG8

Subduction of cold water and the idea of Ekman hats. This program
is a variation of the above program and is useful for illustrating such
things as Ekman hats. Its main difference is that it does not have any
flux across the interfaces; therefore it is in this respect a purely wind-
driven ocean, in which, however, the Ekman pumping does force
water of differing density downward. You may enter your own choice
of parameters as earlier, instead of the default values. When the display comes on, you will see a single view of the ocean, with a time
scale across the top and a display of the distribution of Ekman pumping with latitude on the right-hand side. Otherwise the screen seems
to be dead. Strike one of the cursor keys, and the top of a little column
of water appears in the ocean. Attached to it is an arrow that indicates
the local velocity strength and direction. Color 1 is the lower layer,
color 2 the upper. There is a latitude of outcrop shown also across the
ocean. If the column first appears north of this line, it is a column of
lower-layer water; otherwise it is upper-layer water. Incidentally, the
color-3 curve across the ocean is the boundary of the shadow zone.

Now strike G (for "go") to continue, and the column begins to move.
If it is a lower-layer column, it actually moves along the path of a real
lower-layer column, and its bottom does not change depth. With no interfacial flux, the characteristic curves and contours of constant D are
the same, so we can follow a column along a characteristic. This is not
true with interfacial fluxes, as in the earlier models. In point of his-
tory, this program with no interfacial flux antedates the others. A section showing the total water column along this path is displayed on its
side on the right. You will be surprised to see a little top layer of color

8. This is the accumulated Ekman-hat water from the Ekman pumping; the original color-1 part of the lower-layer column shrinks as it moves southward according to Rule 3. Since its lower end stays at a fixed depth, and it shrinks, you can see how necessary the hat is to keep the whole column at constant height. Hats on for Ekman!

Eventually the column reaches the subduction latitude, where it must sink under a top color-2 layer. Watch it as it slips under the outcrop, Ekman hat and all. As this lower-layer column keeps moving farther southward (and westward), it shrinks even more (color-8 Ekman hat and original color-2 column obeying Rule 3 together). Now, it is the increasing depth of the top color-8 layer that enables it to do so. Remember that we are not following the columns moving in the upper layer, only those in the lower layer. They move in different directions. So on the section developing on the right, the color-2 portion only shows the local depth.

Refer to the text for more about the Ekman hat industry.

PROG9

This program shows a key that tells you what color your computer chooses to correspond to the color numbers referred to in the text.

The Programs

```
1 'PROG0 **
2 KEY OFF: SCREEN 2: CLS
4 LOCATE 1,5: PRINT "A VIEW OF THE SEA"
6 LOCATE 3,5: PRINT "COPYRIGHT PRINCETON UNIVERSITY PRESS, 1986"
8 LOCATE 5,10: PRINT "CONTENTS:"
10 LOCATE 7,10: PRINT "PROGRAM 1: FORCES ON PENDULUM GOING AROUND AN AXIS"
11 LOCATE 8,10: PRINT "AND GRAVITATING TOWARD THE CENTER"
12 LOCATE 9 ,10:PRINT "PROGRAM 2: TWO VIEWS OF A PARTICLE SLIDING ON AN ELLIPSOI
DAL EARTH"
13 LOCATE 12,10: PRINT "PROGRAM 5: FOUR INTERFACE DIAGNOSTICS"
14 LOCATE 10,10:PRINT "PROGRAM 3: BETA-TRIANGLE CRUISE."
15 LOCATE 13,10: PRINT "PROGRAM 6: TUNING FOR CORRECT LATITUDE AND SPEED "
16 LOCATE 11,10:PRINT "PROGRAM 4: DIAGNOSTICS OF A BLOCK OF OCEAN."
18 LOCATE 14,10:PRINT "PROGRAM 7: EXAMPLE OF A MODEL COMPUTED BY CHARACTERISTICS
"
20 LOCATE 15,10: PRINT "PROGRAM 8: SUBDUCTION OF COLD WATER AND EKMAN HATS"
21 LOCATE 16,10: PRINT "PROGRAM 9: COLOR CODE"
22 LOCATE 20,10: INPUT " WHICH PROGRAM (1-9)  "; PROG
24 IF PROG = 1 THEN LOAD "PROG1",R
26 IF PROG = 2 THEN LOAD"PROG2",R
28 IF PROG = 3 THEN LOAD"PROG3",R
30 IF PROG = 4 THEN LOAD"PROG4",R
32 IF PROG = 5 THEN LOAD"PROG5",R
34 IF PROG = 6 THEN LOAD"PROG6",R
36 IF PROG = 7 THEN LOAD"PROG7",R
37 IF PROG = 8 THEN LOAD"PROG8",R
38 IF PROG = 9 THEN LOAD"PROG9",R
```

PROG1

```
1000 'PROG1 **
1001 KEY OFF:SCREEN 1:COLOR 0,2:CLS
1002 LOCATE 1,1:INPUT "SPIN ( 0 - 1 ) = "; SPIN: S = 1/(1-(SPIN^2))
1004 IF SPIN > .9999 THEN CLS: LOCATE 10,20 : IF SPIN >.9999 THEN PRINT" GLOBE F
LIES APART"
1006 IF SPIN > .9999 THEN GOTO 1004
1008 B = 100/(S^(1/3))
1010 A = B*SQR(S)
1012 X = (A*B)*SQR(1/(A*A+B*B)): Y = X
1014 CIRCLE(150,100),A            ,1,0,3.1415*2,SQR(1/S)
1016 LOCATE 12,22: PRINT "L"
1018 LINE (150-A,100)-(150+A,100),1
1020 LINE (150,100+B)-(150,100-B),1
1022 LINE (150,100)-(150+10,100-10),1
1024 T = ATN(S): XM =X- 20*COS(T):YM=X-20*SIN(T)
1026 LINE (150+XM,100-YM)-(150+XM-X+YM,100-2*YM+Y ),2
1028 LINE(150+X,100-X)-(150+XM        ,100-YM      ),3
1030 LINE(150+YM,100-YM)-(150+XM,100-YM),8
1032 CIRCLE(150+XM,100-YM),3,8
1034 PAINT (150+XM,100-YM),8
1036 GOTO 1002
```

PROG2

```
2000 'PROG2
2002 KEY OFF: SCREEN 1: COLOR 0,2: CLS
2004 'ELLIPSOIDAL EARTH
2006 'ABSOLUTE 1 AND 2, RELATIVE 3 AND 4; EXPERT ODD, NOVICE EVEN
2008 PRINT "TWO VIEWS" :PRINT"START ":PRINT"PARTICLE"
2010 LOCATE 2,30:PRINT"ON BACK "
2012 CIRCLE(292,12),1,1
2014 LOCATE 21,6:PRINT"ABSOLUTE"
2016 LOCATE 21,24:PRINT"RELATIVE"
2018 CIRCLE(80,12),2,2:CIRCLE(80,20),1,3
2020 PI = 3.14159: W=10: W2 = W^2 : DT =.005: FACT = 60:INC=PI/16
2022 GOSUB 2074
2024 LOCATE 23,1 :INPUT "LAT";LA
2026 FOR I = 1 TO 4: LA(I)=LA*PI/180: LO(I)=-PI/2      :NEXT
2028 LAD(1)=0: LAD(3)=0 :LAD(2)=V:LAD(4)=LAD(2)
2030 LOCATE 23,20: INPUT "U,V";U,V
2032 LAD(1)=0: LAD(3)=0 :LAD(2)=V:LAD(4)=LAD(2)
2034 LOD(2)=U/COS(LA(1))+W : LOD(4)=LOD(2)
```

142

```
2036 LODD(2)=2*TAN(LA(2))*LAD(2)*LOD(2)
2038 LADD(2)=SIN(LA(2))*COS(LA(2))*(W2-(LOD(2))^2)
2040 LOD(2)=LOD(2)+DT*LODD(2):LAD(2)=LAD(2)+LADD(2)*DT
2042 T = T+DT
2044 LA(2)=LA(2)+DT*LAD(2)    : LO(2)=LO(2)+DT*LOD(2): LO(4)=LO(2)-W*T
2046 LA(4)=LA(2): LO(1)=LO(1)+W*DT
2048 FOR I = 1 TO 4: RHO(I)=COS(LA(I))
2050 X(I)=RHO(I)*COS(LO(I)):Y(I)=RHO(I)*SIN(LO(I))*SIN(INC)
2052 Z(I)=Y(I)+SIN(LA(I))*COS(INC)
2054 IF Y(1)<0 THEN C(1)=2 ELSE C(1)=1
2056 IF Y(2)<0 THEN C(2)=3 ELSE C(2)=1
2058 IF Y(3)<0 THEN C(3)=2 ELSE C(3)=1
2060 IF Y(4)<0 THEN C(4)=3 ELSE C(4)=1
2062 IF I = 1 THEN CIRCLE(70+FACT*X(1),90-FACT*Z(1)),2,C(1)
2064 IF I = 2 THEN CIRCLE(70+FACT*X(2),90-FACT*Z(2)),1,C(2)
2066 IF I = 3 THEN CIRCLE(220+FACT*X(3),90-FACT*Z(3)),2,C(3)
2068 IF I = 4 THEN CIRCLE(220+FACT*X(4),90-FACT*Z(4)),1,C(4)
2070 NEXT
2072 GOTO 2036
2074 J = INC: ' PLOT THE TWO SPHERES "
2076 FOR I = 0 TO 1
2078 CIRCLE (220-150*I,90),60,1,0,2*PI,1
2080 CIRCLE (220-150*I,90),60,1,0,2*PI,SIN(J)
2082 C9=60*SIN(PI/6)*COS(J)
2084 K9=60*COS(PI/6)
2086 CIRCLE (220-150*I,90-C9),K9,1,0,2*PI,SIN(J)
2088 CIRCLE (220-150*I,90+C9),K9,1,0,2*PI,SIN(J)
2090 C9=60*SIN(PI/3)*COS(J)
2092 K9=60*COS(PI/3)
2094 CIRCLE(220-150*I,90-C9),K9,1,0,2*PI,SIN(J)
2096 CIRCLE(220-150*I,90+C9),K9,1,0,2*PI,SIN(J)
2098 NEXT
2100 RETURN
```

PROG3

```
3000 ' PROG3
3002 SCREEN 1:PRINT" BETA SPIRAL CRUISF  "
3004 PRINT " " :PRINT" PRESS C TO CONTINUE"
3006 PRINT"       H TO CHANGE HEADING"
3008 PRINT"       M FOR MARKER"
3010 PRINT"       S FOR CTD STATION"
3012 B$=INKEY$
3014 ZZ - RND
3016 IF B$="C" THEN GOTO 3018 ELSE GOTO 3012
3018 KEY OFF: SCREEN 1: COLOR 0,2: CLS
3020 XC = 245:YC=62: PI=3.14159
3022 FOR N = 1 TO 6:FOR M = 1 TO 10
3024  SPEED = 1 : DT =.1 : DIR = 210
3026 DIR = DIR*3.1415/180
3028  MS = 200*RND
3030 LINE(30*M,30*N)-(0,0),1,B
3032 NEXT:NEXT
3034 DRAW " BM+243,0c2TA160U30TA250U20TA0D15L8D12R2D3
3036 DRAW " TA120U30TA160U30TA180U24TA230U30TA270U30TA280U10TA230U10TA180U24
3038 DRAW " BM+230, 180TA235U40TA225U20TA235U20TA180U10TA90U20TA100U30TA130U50TA
     190U15TA100U4TA280U15TA80U15TA180U10
3040 LOCATE 23,1:PRINT "90W"
3042 LOCATE 23,34:PRINT "0"
3044 DRAW " TA180 BM 0,118TA270U10TA170U10TA270U10TA270U 1TA180U20TA240U30TA290U
     30TA250U20TA240U15TA220U18
3046 LOCATE 1,38:PRINT "60N"
3048 LOCATE 23,40:PRINT "0";
3050 DRAW"BM20,119TA280U7TA250U14TA180U3TA70U16TA100U5TA0U3
3052 PSET(45,125),2
3054 DRAW" TA300U5TA240U5TA120U5TA60U5
3056 PSET ( 88,135),2
3058 PSET ( 89,140),2
3060 PSET ( 88,145),2
3062 LINE(0,0)-(300,180),2,B
3064 PAINT(1,1),2
3066 PAINT (299,1),2
3068 PAINT(1,179),2
3070 PAINT (25,119),2
3072 PAINT (49,125),2
3074 LINE(72,259)-(70,259),2
3076 PSET(78,84),2
3078 PSET(186,66),2
```

143

```
3080 PSET(186,64.5),2
3082 PSET(198,67.5),2
3084 PSET(219,81),2
3086 PSET(222,99),2
3088 PSET(198,132),2: PSET(206,94),2
3090 PSET(186,114),1: X = 204: Y = 96
3092 DRAW" C1TA10U30TA130U30TA250U30
3094 LINE(0,0)-(300,60),0,BF
3096 LINE(0,0)-(300,60),2,B
3098 LOCATE 3,8 : PRINT "HYDROGRAPHIC SECTION"
3100 CIRCLE(206,94),2,0:PAINT(206,94),0
3102 CIRCLE(245,62),2,0:PAINT(245,62),0
3104 ' INTEGRATION OF SHIP'S PATH
3106 XD=SPEED*SIN(DIR):YD=-SPEED*COS(DIR)
3108 X = DT*XD+X: Y=YD*DT+Y
3110 PSET(X,Y),2
3112 LINE(300-M,1)-(300-M,59),0: LINE(299-M ,0)-(299-M ,60),1
3114 MD = SQR(XD^2+YD^2)
3116 M =2*DT*MD+M
3118 IF FLAG = 1 THEN GOTO 3122
3120 IF M >    MS THEN GOSUB 3184
3122 ML = M
3124 A$ = INKEY$
3126 DL = INT(21-M/25-NUM/12-STORM)
3128 CIRCLE(206,94),2,3:PAINT(206,94),3
3130 CIRCLE(245,62),2,3:PAINT(245,62),3
3132 LOCATE 21,1: PRINT"STATION NO.";INT(NUM)
3134 LOCATE 21,19: PRINT"DAYS LEFT=";DL
3136 LOCATE 22,20
3138 IF A$ = "H" THEN INPUT "H= ";DIR : DIR = DIR*(3.1415)/180
3140 IF A$ = "S" THEN GOSUB 3152
3142 IF A$ = "M" THEN GOSUB 3178
3144 S1 =12.5
3146 RC = SQR((XC-X)^2+(YC-Y)^2)
3148 IF RC>DL*S1 THEN GOTO 3202
3150 GOTO 3100
3152 FOR I = 1 TO 5
3154 P(I)= (I-1)*PI/10
3156 H0(I)=5*I: H1(I) = 4*I+20
3158 ANG =((Y-110)/30)*PI/2 + P(I)
3160 H(I)=H0(I)+(1-X/300)*H1(I)*COS(ANG)
3162 IF NUM = 0 THEN HL(I)=H(I)
3164 IF NUM = 0 THEN N=M-1
3166 LINE(300-M,H(I))-(300-N ,HL(I)),1+I MOD 3
3168 HL(I)=H(I)
3170 NEXT
3172 NUM = NUM+1: N = M
3174 CIRCLE(X,Y),1,3
3176 RETURN
3178 LINE(301-M,0)-(301-M,60),1
3180 LINE(X+1,Y+1)-(X-1,Y-1),1,B
3182 RETURN
3184 LOCATE 9,2 :PRINT"THREE DAY N.E. GALE"
3186 FOR I = 1 TO 30
3188 XS=X:YS=Y
3190 X = X-.4:Y=Y+.3
3192 PSET(X,Y),2
3194 BEEP      : NEXT
3196 STORM = 3
3198 FLAG = 1
3200 RETURN
3202 XO = X ; YO = Y
3204 LOCATE 9,2: PRINT "CAPTAIN ORDERS SHIP TO CADIZ"
3206 SOUND 200,40:BEEP:BEEP:BEEP
3208 XD = SPEED*(XC-XO)/RC:YD= SPEED*(YC-YO)/RC
3210 X = XD*DT+X: Y = YD*DT+Y
3212 PSET (X,Y),2
3214 IF X>XC THEN STOP
3216 M = 2*DT*MD+M
3218 DL = INT(21-M/25-NUM/12-STORM)
3220 LOCATE 21,19:PRINT"DAYS LEFT=";DL
3222 GOTO 3208
```

PROGRAM 4

PROG4

```
4000 'PROG4        **
4001 SCREEN 1   : COLOR 0,2 : KEY OFF: CLS
4004  H0=300:H1=300:H2=300
4006  D1=800:D2=800:D0=800
4008  B0=4000:B1=4000:B2=4000
4014 GOSUB 4146                    'DRAW BLOCK
4016 GOSUB 4192:GOSUB 4220:GOSUB 4248        'DRAW INTERFACES
4018 PRINT "DIAGNOSTIC WITH 3 RULES"
4019 LOCATE 1,29:PRINT"N"
4020 LOCATE 2,36:PRINT"E"
4021 LOCATE 5,38:PRINT "1"
4022 LOCATE 6,38: PRINT "2"
4023 LOCATE 2,1
4024 INPUT "SIGMA T DIFFS G1,G2";G3,G4
4026 PRINT " ENTER INTERFACE HEIGHTS"
4028 PRINT " NW,SW,SE CORNERS METERS"
4032 LOCATE 5,1
4033 H00=H0:H01=H1:H02=H2:INPUT "INTERFACE 1";H1,H0,H2 : GOSUB 4276:GOSUB 4146
4034 GOSUB 4192
4036 D00=D0:D01=D1:D02=D2:INPUT "INTERFACE 2";D1,D0,D2 : GOSUB 4304:GOSUB 4146
4038 GOSUB 4220
4040 B00=B0:B01=B1:B02=B2:INPUT "BOTTOM        ";B1,B0,B2 : GOSUB 4332:GOSUB 4146
4042 GOSUB 4248
4044 INPUT "LATITUDE=";L
4046 FIRST=0
4048 L1 = ((L-5)/90)*(3.1415)/2
4050 L2 = ((L+5)/90)*(3.1415)/2
4052 F0 = 1.4584E-04 *SIN(L1)
4054 F1 = 1.4584E-04 *SIN(L2)
4055 FM = .5*F1+.5*F0: G1=G3/FM: G2=G4/FM: 'CORIOLIS PARAMETERS
4056      GOSUB 4146:GOSUB 4192:GOSUB 4220:GOSUB 4248
4058 A0 = (B0-D0)/F0   :'POTENTIAL THICKNESSES
4060 A1 = (B1-D1)/F1
4062 A2 = (B2-D2)/F0
4064 C0 = (D0-H0)/F0
4066 C1 = (D1-H1)/F1
4068 C2 = (D2-H2)/F0
4070 E0 = H0/F0
4072 E1 = H1/F1
4074 E2 = H2/F0
4076 X = 1110000!  * COS((L /90)*3.1415/2)
4078 Y = 1110000!
4080 LOCATE 23,1:INPUT"V CM/SEC,T DIR TRUE";V,T8 : T=360-T8:
4082 LOCATE 19,10: PRINT "GV CM/SEC":LOCATE 16,18:PRINT"WE W1 W2 "
4084 IF FIRST>0 THEN GOSUB 4134
4086 FIRST=1
4088 T1 = (T/90)*3.1415/2
4090 P1 = V*COS(T1); P2=V*SIN(T1)   :' BOTTOM LAYER VELOCITY,CM/SEC
4094 W2 = P2*(A2-A0)/X-P1*(A1-A0)/Y :' BOTTOM LAYER DIVERGENCE
4096 Q1 = P1 +G2* (D2-D0)/X : ' FOR OTHER LAYERS
4098 Q2 = P2 +G2* (D1-D0)/Y
4100 R1 = Q1 +G1*(H2-H0)/X
4102 R2 = Q2 +G1*(H1-H0)/Y
4104 W3 = Q2*(C2-C0)/X-Q1*(C1-C0)/Y
4106 W4 = R2*(E2-E0)/X-R1*(E1-E0)/Y
4108 W1 = W2+W3: SP=86400!*FM: 'FLUXES ACROSS INTERFACES, CM/DAY
4110 W0 = W2+W3+W4: W0=W0*SP:W1=W1*SP:W2=W2*SP: ' W0 IS EKMAN
4112 LOCATE 22,1:PRINT USING "WE=###.## W1=###.## W2=###.## CM/DAY";W0,W1,W2:'**
4114 CIRCLE (100,100),20,1,0,6,2,1: '1 CM/SEC CIRCLE
4116 LINE(100,100)- (-20*P2+100,100-20*P1),1 ' PLOTS GV'S
4118 LINE( 100,100)-(100-20*R2,100-20*R1),2
4120 LINE(100,100)-(100-20*Q2,100-20*Q1),3
4122 IF I=0 THEN LINE (150,100)-(210,100),1
4124 I=I+2
4126 LINE (150+I,100)-(150+I,100-    W0),7 :'PLOTS THE W'S
4128 LINE (170+I,100)-(170+I,100-    W1),2
4130 LINE (190+I,100)-(190+I,100-    W2),3
4132 GOTO 4080
4134 'REMOVE OLD LINES
4136 LINE(100,100)- (-20*P2+100,100-20*P1),0
4138 LINE( 100,100)-(100-20*R2,100-20*R1),0
4140 LINE(100,100)-(100-20*Q2,100-20*Q1),0
4142 PSET (-20*P2+100,100-20*P1),3
4144 RETURN
4146 'DRAW BLOCK
4148 N=120
```

145

```
4150 LINE (220,20  )-(220,20+N),1
4152 LINE (260,50  )-(260,50+N),1
4154 LINE (290,30  )-(290,30+N),1
4156 FOR N=0 TO 120 STEP 5
4158 PSET (250,N),1
4160 NEXT
4162 N =  30
4164 LINE(220,20)-(220+N*4/3,20+N),1
4166 LINE(250, 0)-(250+N*4/3 ,N),1
4168 LINE(220,20)-(220+N,20-2*N/3),1
4170 LINE (260,50)- (260 +N,50-(2/3)*N),1
4172 LINE (220,140)-(220 +N*(4/3),140+N),1
4174 LINE (260,170)-(260 +N,170-(2/3)*N),1
4176 FOR N =0 TO 30 STEP 5
4178 PSET (250 +N*(4/3),N+120),1
4180 PSET (220+N,140-(2/3)*N),1
4182 NEXT
4184 FOR N=0 TO 5
4186 PSET (218,N*25+20),1
4188 NEXT
4190 RETURN
4192 'INTERFACE 1
4194 H3 = (H1+H2)/2
4196 H4=2*H3-H0
4198 N = 40
4200 LINE (220,20+H1/40)-(220+N,20+H1/40+(3/4+(H0-H1)/1600)*N),2
4202 FOR N = 0 TO 40 STEP 3
4204 PSET (250+N,H4/40+(3/4+(H0-H1)/1600)*N),2
4206 NEXT
4208 N=30
4210 LINE (260,50+H0/40)-(260+N,50+H0/40-(2/3-(H2-H0)/1200)*N),2
4212 FOR N = 0 TO 30 STEP 3
4214 PSET (220+N,20+H1/40-(2/3-(H2-H0)/1200)*N),2
4216 NEXT
4218 RETURN
4220 'INTERFACE 2
4222 D3 = (D1+D2)/2
4224 D4=2*D3-D0
4226 N = 40
4228 LINE (220,20+D1/40)-(220+N,20+D1/40+(3/4+(D0-D1)/1600)*N),3
4230 FOR N = 0 TO 40 STEP 3
4232 PSET (250+N,D4/40+(3/4+(D0-D1)/1600)*N),3
4234 NEXT
4236 N=30
4238 LINE (260,50+D0/40)-(260+N,50+D0/40-(2/3-(D2-D0)/1200)*N),3
4240 FOR N = 0 TO 30 STEP 3
4242 PSET (220+N,20+D1/40-(2/3-(D2-D0)/1200)*N),3
4244 NEXT
4246 RETURN
4248 'BOTTOM
4250 B3 = (B1+B2)/2
4252 B4=2*B3-B0
4254 N = 40
4256 LINE (220,20+B1/40)-(220+N,20+B1/40+(3/4+(B0-B1)/1600)*N),1
4258 FOR N = 0 TO 40 STEP 3
4260 PSET (250+N,B4/40+(3/4+(B0-B1)/1600)*N),1
4262 NEXT
4264 N=30
4266 LINE (260,50+B0/40)-(260+N,50+B0/40-(2/3-(B2-B0)/1200)*N),1
4268 FOR N = 0 TO 30 STEP 3
4270 PSET (220+N,20+B1/40-(2/3-(B2-B0)/1200)*N),1
4272 NEXT
4274 RETURN
4276 ' OLD INTERFACE 1
4278 HO3 = (HO1+HO2)/2
4280 HO4=2*HO3-HO0
4282 N = 40
4284 LINE (220,20+HO1/40)-(220+N,20+HO1/40+(3/4+(HO0-HO1)/1600)*N),0
4286 FOR N = 0 TO 40 STEP 3
4288 PSET (250+N,HO4/40+(3/4+(HO0-HO1)/1600)*N),0
4290 NEXT
4292 N=30
4294 LINE (260,50+HO0/40)-(260+N,50+HO0/40-(2/3-(HO2-HO0)/1200)*N),0
4296 FOR N = 0 TO 30 STEP 3
4298 PSET (220+N,20+HO1/40-(2/3-(HO2-HO0)/1200)*N),0
4300 NEXT
4302 RETURN
4304 'OLD INTERFACE 2
4306 DO3 = (DO1+DO2)/2
```

```
4308 DO4=2*DO3-DOØ
4310 N = 40
4312 LINE (220,20+DO1/40)-(220+N,20+DO1/40+(3/4+(DOØ-DO1)/1600)*N),Ø
4314 FOR N = Ø TO 40 STEP 3
4316 PSET (250+N,DO4/40+(3/4+(DOØ-DO1)/1600)*N),Ø
4318 NEXT
4320 N=30
4322 LINE (260,50+DOØ/40)-(260+N,50+DOØ/40-(2/3-(DO2-DOØ)/1200)*N),Ø
4324 FOR N = Ø TO 30 STEP 3
4326 PSET (220+N,20+DO1/40-(2/3-(DO2-DOØ)/1200)*N),Ø
4328 NEXT
4330 RETURN
4332 ' OLD BOTTOM
4334 BO3 = (BO1+BO2)/2
4336 BO4=2*BO3-BOØ
4338 N = 40
4340 LINE (220,20+BO1/40)-(220+N,20+BO1/40+(3/4+(BOØ-BO1)/1600)*N),Ø
4342 FOR N = Ø TO 40 STEP 3
4344 PSET (250+N,BO4/40+(3/4+(BOØ-BO1)/1600)*N),Ø
4346 NEXT
4348 N=30
4350 LINE (260,50+BOØ/40)-(260+N,50+BOØ/40-(2/3-(BO2-BOØ)/1200)*N),Ø
4352 FOR N = Ø TO 30 STEP 3
4354 PSET (220+N,20+BO1/40-(2/3-(BO2-BOØ)/1200)*N),Ø
4356 NEXT
4358 RETURN
```

PROG5

```
5000 'PROG5  **
5002 SCREEN 2
5004 SØ=100:S1=100:S2=100
5006 HØ=300:H1=300:H2=300
5008 D1=800:D2=800:DØ=800
5010 BØ=4000:B1=4000:B2=4000
5012 CLS
5014 SCREEN 1  : COLOR Ø,2:KEY OFF
5016 GOSUB 5162                     'DRAW BLOCK
5017 LOCATE 1,29:PRINT "N":LOCATE 2,36: PRINT "E"
5018 GOSUB 5208:GOSUB 5236:GOSUB 5264:GOSUB 5292         'DRAW INTERFACES
5020 PRINT "THREE RULE MODEL"
5022 INPUT "G1,G2,G3";G4,G5,G6
5030 PRINT " ENTER INTERFACE HEIGHTS"
5032 PRINT " NW,SW,SE CORNER METERS"
5036 SOØ=SØ:SO1=S1:SO2=S2:INPUT "1: "  ;S1,SØ,S2 : GOSUB 5320:GOSUB 5162
5038 GOSUB 5208
5040 HOØ=HØ:HO1=H1:HO2=H2:INPUT "2: ";H1,HØ,H2 : GOSUB 5348:GOSUB 5162
5042 GOSUB 5236
5044 DOØ=DØ:DO1=D1:DO2=D2:INPUT "3: ";D1,DØ,D2 : GOSUB 5376:GOSUB 5162
5046 GOSUB 5264
5048 BOØ=BØ:BO1=B1:BO2=B2:INPUT "4: ";B1,BØ,B2 : GOSUB 5404:GOSUB 5162
5050 GOSUB 5292
5052 INPUT "LATITUDE=";L
5064 FIRST=Ø
5066 L1 = ((L-5)/90)*(3.1415)/2
5068 L2 = ((L+5)/90)*(3.1415)/2
5070 FØ = 1.4584E-04  *SIN(L1)
5072 F1 = 1.4584E-04  *SIN(L2) : FM = .5*F1+.5*FØ
5073 G1=G4/FM:G2=G5/FM:G3=G6/FM
5074 LOCATE 22,1
5076 CLS: GOSUB 5162:GOSUB 5208:GOSUB 5236:GOSUB 5264:GOSUB 5292
5078 AØ = (BØ-DØ)/FØ        :'POTENTIAL THICKNESSES
5080 A1 = (B1-D1)/F1
5082 A2 = (B2-D2)/FØ
5084 CØ = (DØ-HØ)/FØ     :JØ=SØ/FØ
5086 C1 = (D1-H1)/F1    : J1=S1/F1: J2=S2/FØ
5088 C2 = (D2-H2)/FØ
5090 EØ = (HØ-SØ)/FØ
5092 E1 = (H1-S1)/F1
5094 E2 = (H2-S2)/FØ
5096 X = 1110000!  * COS((L /90)*3.1415/2):'ARC LENGTHS OF BOX, METERS
5098 Y = 1110000!
5100 LOCATE 1,1: INPUT"V CM/SEC,T TRUE";V,T8 : T=360-T8
5101 LINE (0,20)-(120,150),Ø,BF
5102 IF FIRST>Ø THEN GOSUB 5150
5103 LOCATE 17,1: PRINT "GV'S CM/SEC"
5104 FIRST=1
```

```
5106 T1 = (T/90)*3.1415/2
5108 P1 = V*COS(T1)      :'VELOCITIES CM/SEC BOTTOM LAYER
5110 P2 = V*SIN(T1)
5114 Q1 = P1 +G3* (D2-D0)/X  : ' HORIZ VELOCITIES
5116 Q2 = P2 +G3* (D1-D0)/Y
5118 R1 = Q1 +G2*(H2-H0)/X   : Z1=R1+G1*(S2-S0)/X
5120 R2 = Q2 +G2*(H1-H0)/Y   : Z2=R2+G1*(S1-S0)/Y
5121 W3 = P2*(A2-A0)/X-P1*(A1-A0)/Y   :' LAYER DIVERGENCES
5122 W4 = Q2*(C2-C0)/X-Q1*(C1-C0)/Y
5124 W5 = R2*(E2-E0)/X-R1*(E1-E0)/Y  : W6= Z2*(J2-J0)/X-Z1*(J1-J0)/Y
5126 W3 = W3: W2=W3+W4 :W1=W3+W4+W5 :WE=W3+W4+W5+W6:       SP=86400!*FM
5128 W1=SP*W1:W2=SP*W2:W3=W3*SP:WE=SP*WE:'INTERFACIAL FLUXES
5129 LOCATE 20,1: PRINT USING "WE ###.### W1 ###.### ";WE,W1 : LOCATE 21,1:PRINT
    USING "W2 ###.### W3 ###.### CM/DAY";W2,W3
5130 CIRCLE ( 70,100),20,1,0,6,2,1
5132 LINE( 70,100)- (-20*P2+ 70,100-20*P1),1  :'DRAWS GV'S CM/SEC
5134 LINE( 70,100)-( 70-20*R2,100-20*R1),2
5136 LINE( 70,100)-( 70-20*Q2,100-20*Q1),3
5137 LINE( 70,100)-( 70-20*Z2,100-20*Z1),1
5138 FOR KK=0 TO 75  STEP 6 : PSET(130+KK,100),1: NEXT
5140 I=I+2
5141 LINE (130+I,100)-(130+I,100-    WE),1: ' PLOTS THE W'S
5142 LINE (155+I,100)-(155+I,100-    W1),2: ' PLOTS THE W'S
5144 LINE (180+I,100)-(180+I,100-    W2),3
5146 LINE (205+I,100)-(205+I,100-    W3),1
5147 LOCATE 15,16: PRINT " WE W1 W2 W3"
5148 GOTO 5100
5150 'REMOVE OLD LINES
5152 LINE(100,100)- (-20*P2+100,100-20*P1),0
5154 LINE( 100,100)-(100-20*R2,100-20*R1),0
5156 LINE(100,100)-(100-20*Q2,100-20*Q1),0
5158 'PSET (-20*P2+100,100-20*P1),3
5160 RETURN
5162 'DRAW BLOCK
5164 N=120
5166 LINE (220,20  )-(220,20+N),1
5168 LINE (260,50  )-(260,50+N),1
5170 LINE (290,30  )-(290,30+N),1
5172 FOR N=0 TO 120 STEP 5
5174 PSET (250,N),1
5176 NEXT
5178 N =  30
5180 LINE(220,20)-(220+N*4/3,20+N),1
5182 LINE(250, 0)-(250+N*4/3  ,N),1
5184 LINE(220,20)-(220+N,20-2*N/3),1
5186 LINE (260,50)- (260 +N,50-(2/3)*N),1
5188 LINE (220,140)-(220 +N*(4/3),140+N),1
5190 LINE (260,170)-(260 +N,170-(2/3)*N),1
5192 FOR N =0 TO 30 STEP 5
5194 PSET (250 +N*(4/3),N+120),1
5196 PSET (220+N,140-(2/3)*N),1
5198 NEXT
5200 FOR N=0 TO 5
5202 PSET (218,N*25+20),1
5204 NEXT
5206 RETURN
5208 'INTERFACE 1
5210 S3 = (S1+S2)/2
5212 S4=2*S3-S0
5214 N=40
5216 LINE (220,20+S1/40)-(220+N,20+S1/40+(3/4+(S0-S1)/1600)*N),1
5218 FOR N = 0 TO 40 STEP 3
5220 PSET (250+N,S4/40+(3/4+(S0-S1)/1600)*N),1
5222 NEXT
5224 N=30
5226 LINE (260,50+S0/40)-(260+N,50+S0/40-(2/3-(S2-S0)/1200)*N),1
5228 FOR N = 0 TO 30 STEP 3
5230 PSET (220+N,20+S1/40-(2/3-(S2-S0)/1200)*N),1
5232 NEXT
5234 RETURN
5236 'INTERFACE 1
5238 H3 = (H1+H2)/2
5240 H4=2*H3-H0
5242 N = 40
5244 LINE (220,20+H1/40)-(220+N,20+H1/40+(3/4+(H0-H1)/1600)*N),2
5246 FOR N = 0 TO 40 STEP 3
5248 PSET (250+N,H4/40+(3/4+(H0-H1)/1600)*N),2
5250 NEXT
5252 N=30
```

```
5254 LINE (260,50+HØ/40)-(260+N,50+HØ/40-(2/3-(H2-HØ)/1200)*N),2
5256 FOR N = Ø TO 30 STEP 3
5258 PSET (220+N,20+H1/40-(2/3-(H2-HØ)/1200)*N),2
5260 NEXT
5262 RETURN
5264 'INTERFACE 3
5266 D3 = (D1+D2)/2
5268 D4=2*D3-DØ
5270 N = 40
5272 LINE (220,20+D1/40)-(220+N,20+D1/40+(3/4+(DØ-D1)/1600)*N),3
5274 FOR N = Ø TO 40 STEP 3
5276 PSET (250+N,D4/40+(3/4+(DØ-D1)/1600)*N),3
5278 NEXT
5280 N=30
5282 LINE (260,50+DØ/40)-(260+N,50+DØ/40-(2/3-(D2-DØ)/1200)*N),3
5284 FOR N = Ø TO 30 STEP 3
5286 PSET (220+N,20+D1/40-(2/3-(D2-DØ)/1200)*N),3
5288 NEXT
5290 RETURN
5292 'BOTTOM
5294 B3 = (B1+B2)/2
5296 B4=2*B3-BØ
5298 N = 40
5300 LINE (220,20+B1/40)-(220+N,20+B1/40+(3/4+(BØ-B1)/1600)*N),1
5302 FOR N = Ø TO 40 STEP 3
5304 PSET (250+N,B4/40+(3/4+(BØ-B1)/1600)*N),1
5306 NEXT
5308 N=30
5310 LINE (260,50+BØ/40)-(260+N,50+BØ/40-(2/3-(B2-BØ)/1200)*N),1
5312 FOR N = Ø TO 30 STEP 3
5314 PSET (220+N,20+B1/40-(2/3-(B2-BØ)/1200)*N),1
5316 NEXT
5318 RETURN
5320 ' OLD INTERFACE 1
5322 SO3 = (SO1+SO2)/2
5324 SO4=2*SO3-SOØ
5326 N = 40
5328 LINE (220,20+SO1/40)-(220+N,20+SO1/40+(3/4+(SOØ-SO1)/1600)*N),Ø
5330 FOR N = Ø TO 40 STEP 3
5332 PSET (250+N,SO4/40+(3/4+(SOØ-SO1)/1600)*N),Ø
5334 NEXT
5336 N=30
5338 LINE (260,50+SOØ/40)-(260+N,50+SOØ/40-(2/3-(SO2-SOØ)/1200)*N),Ø
5340 FOR N = Ø TO 30 STEP 3
5342 PSET (220+N,20+SO1/40-(2/3-(SO2-SOØ)/1200)*N),Ø
5344 NEXT
5346 RETURN
5348 ' OLD INTERFACE 2
5350 HO3 = (HO1+HO2)/2
5352 HO4=2*HO3-HOØ
5354 N = 40
5356 LINE (220,20+HO1/40)-(220+N,20+HO1/40+(3/4+(HOØ-HO1)/1600)*N),Ø
5358 FOR N = Ø TO 40 STEP 3
5360 PSET (250+N,HO4/40+(3/4+(HOØ-HO1)/1600)*N),Ø
5362 NEXT
5364 N=30
5366 LINE (260,50+HOØ/40)-(260+N,50+HOØ/40-(2/3-(HO2-HOØ)/1200)*N),Ø
5368 FOR N = Ø TO 30 STEP 3
5370 PSET (220+N,20+HO1/40-(2/3-(HO2-HOØ)/1200)*N),Ø
5372 NEXT
5374 RETURN
5376 'OLD INTERFACE 3
5378 DO3 = (DO1+DO2)/2
5380 DO4=2*DO3-DOØ
5382 N = 40
5384 LINE (220,20+DO1/40)-(220+N,20+DO1/40+(3/4+(DOØ-DO1)/1600)*N),Ø
5386 FOR N = Ø TO 40 STEP 3
5388 PSET (250+N,DO4/40+(3/4+(DOØ-DO1)/1600)*N),Ø
5390 NEXT
5392 N=30
5394 LINE (260,50+DOØ/40)-(260+N,50+DOØ/40-(2/3-(DO2-DOØ)/1200)*N),Ø
5396 FOR N = Ø TO 30 STEP 3
5398 PSET (220+N,20+DO1/40-(2/3-(DO2-DOØ)/1200)*N),Ø
5400 NEXT
5402 RETURN
5404 ' OLD BOTTOM  OR BOTTOM INTERFACE
5406 BO3 = (BO1+BO2)/2
5408 BO4=2*BO3-BOØ
5410 N = 40
```

PROGRAM 6

```
5412 LINE (220,20+B01/40)-(220+N,20+B01/40+(3/4+(B00-B01)/1600)*N),0
5414 FOR N = 0 TO 40 STEP 3
5416 PSET (250+N,B04/40+(3/4+(B00-B01)/1600)*N),0
5418 NEXT
5420 N=30
5422 LINE (260,50+B00/40)-(260+N,50+B00/40-(2/3-(B02-B00)/1200)*N),0
5424 FOR N = 0 TO 30 STEP 3
5426 PSET (220+N,20+B01/40-(2/3-(B02-B00)/1200)*N),0
5428 NEXT
5430 RETURN
```

PROG6

```
6000 'PROG6  **
6002 SCREEN 2    :KEY OFF
6004 S0=340:S1=200:S2=200
6006 H0=400:H1=300:H2=400
6008 D1=800:D2=600:D0=700
6010 B0=800 :B1=1000:B2=700
6012 V = -.6
6014 SCREEN 1  ; COLOR 0,2 :CLS
6016 GOSUB 6130            'DRAW BLOCK
6018 GOSUB 6176:GOSUB 6204:GOSUB 6232:GOSUB 6260      'DRAW INTERFACES
6020 S0=S0/5:S1=S1/5:S2=S2/5
6022 D0=D0/5:D1=D1/5:D2=D2/5
6024 H0=H0/5:H1=H1/5:H2=H2/5
6026 B0=B0/5:B1=B1/5:B2=B2/5
6028 PRINT "CHINESE PUZZLE."
6030 PRINT "RUNS BY ITSELF"
6032 G1=10000
6034 G2 = G1: G3=G1
6036 L=-90
6038 GOTO 6042
6040 GOTO 1250
6042 FIRST=0
6043 LOCATE 15,1: PRINT "SP      EQ      NP"
6044 L1=-1.5
6045 LOCATE 23,13: PRINT"26N"
6046 L2=-1.5+.174528
6048 F0 = 1.4584E-04  *SIN(L1)
6050 F1 = 1.4584E-04  *SIN(L2)
6052 LOCATE 4,1: PRINT USING "v=##.##";V
6054 A0 = (B0-D0)/F0
6056 A1 = (B1-D1)/F1
6058 A2 = (B2-D2)/F0
6060 C0 = (D0-H0)/F0
6062 C1 = (D1-H1)/F1
6064 C2 = (D2-H2)/F0
6066 E0 = (H0-S0)/F0
6068 E1 = (H1-S1)/F1
6070 E2 = (H2-S2)/F0
6072 X = 1110000!  * COS((L /90)*3.1415/2)
6074 Y = 1110000!
6076                  T8=270     : T=360-T8
6078 L1=L1+.04:L2=L2+.04
6080 T1 = (T/90)*3.1415/2
6082 P1 = V*COS(T1)
6084 P2 = V*SIN(T1)
6086 W2 = P2*(A2-A0)/X-P1*(A1-A0)/Y
6088 Q1 = P1 +G2* (D2-D0)/X
6090 Q2 = P2 +G2* (D1-D0)/Y
6092 R1 = Q1 +G1*(H2-H0)/X
6094 R2 = Q2 +G1*(H1-H0)/Y
6096 W3 = Q2*(C2-C0)/X-Q1*(C1-C0)/Y
6098 W4 = R2*(E2-E0)/X-R1*(E1-E0)/Y
6100 W1 = W2+W3
6102 W0 = W2+W3+W4   ; ROLD = R
6104 R2 = W1^2+W2^2+W0^2: R = LOG(R2)-20
6106 NUM=NUM+1: CCC=1+NUM MOD 3
6108 XX = XX+2
6109 IF R>8.8 THEN GOTO 6048
6110 LINE(0,130)-(150,130),1
6116 LINE(XX,130- 5 *ROLD)-(XX+2,130-5*R),CCC
6118 IF L1<.65  AND L2>.7   THEN LINE (XX-2,170)-(XX+2,175),7,BF
6120 IF L2>1.5         THEN V=V+.1 ELSE V = V
6122 IF L2>1.5 THEN XX=0
6124 IF L2>1.5 THEN L1 =-1.5
6126 IF L2>1.5        THEN  L2=-1.5+.174528
6128 GOTO 6048
```

150

```
6130 'DRAW BLOCK
6132 N=120
6134 LINE (220,20  )-(220,20+N),1
6136 LINE (260,50  )-(260,50+N),1
6138 LINE (290,30  )-(290,30+N),1
6140 FOR N=0 TO 120 STEP 5
6142 PSET (250,N),1
6144 NEXT
6146 N = 30
6148 LINE(220,20)-(220+N*4/3,20+N),1
6150 LINE(250, 0)-(250+N*4/3 ,N),1
6152 LINE(220,20)-(220+N,20-2*N/3),1
6154 LINE (260,50)- (260 +N,50-(2/3)*N),1
6156 LINE (220,140)-(220 +N*(4/3),140+N),1
6158 LINE (260,170)-(260 +N,170-(2/3)*N),1
6160 FOR N =0 TO 30 STEP 5
6162 PSET (250 +N*(4/3),N+120),1
6164 PSET (220+N,140-(2/3)*N),1
6166 NEXT
6168 FOR N=0 TO 5
6170 PSET (218,N*25+20),1
6172 NEXT
6174 RETURN
6176 S0=5*S0:S1=5*S1:S2=5*S2:'INTERFACE 0
6178 S3 = (S1+S2)/2
6180 S4=2*S3-S0
6182 N=40
6184 LINE (220,20+S1/40)-(220+N,20+S1/40+(3/4+(S0-S1)/1600)*N),8
6186 FOR N = 0 TO 40 STEP 3
6188 PSET (250+N,S4/40+(3/4+(S0-S1)/1600)*N),8
6190 NEXT
6192 N=30
6194 LINE (260,50+S0/40)-(260+N,50+S0/40-(2/3-(S2-S0)/1200)*N),8
6196 FOR N = 0 TO 30 STEP 3
6198 PSET (220+N,20+S1/40-(2/3-(S2-S0)/1200)*N),8
6200 NEXT
6202 RETURN
6204 H0=5*H0:H1=5*H1:H2=5*H2:'INTERFACE 1
6206 H3 = (H1+H2)/2
6208 H4=2*H3-H0
6210 N = 40
6212 LINE (220,20+H1/40)-(220+N,20+H1/40+(3/4+(H0-H1)/1600)*N),2
6214 FOR N = 0 TO 40 STEP 3
6216 PSET (250+N,H4/40+(3/4+(H0-H1)/1600)*N),2
6218 NEXT
6220 N=30
6222 LINE (260,50+H0/40)-(260+N,50+H0/40-(2/3-(H2-H0)/1200)*N),2
6224 FOR N = 0 TO 30 STEP 3
6226 PSET (220+N,20+H1/40-(2/3-(H2-H0)/1200)*N),2
6228 NEXT
6230 RETURN
6232 D0=5*D0:D1=5*D1:D2=5*D2:'INTERFACE 2
6234 D3 = (D1+D2)/2
6236 D4=2*D3-D0
6238 N = 40
6240 LINE (220,20+D1/40)-(220+N,20+D1/40+(3/4+(D0-D1)/1600)*N),3
6242 FOR N = 0 TO 40 STEP 3
6244 PSET (250+N,D4/40+(3/4+(D0-D1)/1600)*N),3
6246 NEXT
6248 N=30
6250 LINE (260,50+D0/40)-(260+N,50+D0/40-(2/3-(D2-D0)/1200)*N),3
6252 FOR N = 0 TO 30 STEP 3
6254 PSET (220+N,20+D1/40-(2/3-(D2-D0)/1200)*N),3
6256 NEXT
6258 RETURN
6260 B0=5*B0:B1=5*B1:B2=5*B2:'BOTTOM
6262 B3 = (B1+B2)/2
6264 B4=2*B3-B0
6266 N = 40
6268 LINE (220,20+B1/40)-(220+N,20+B1/40+(3/4+(B0-B1)/1600)*N),8
6270 FOR N = 0 TO 40 STEP 3
6272 PSET (250+N,B4/40+(3/4+(B0-B1)/1600)*N),8
6274 NEXT
6276 N=30
6278 LINE (260,50+B0/40)-(260+N,50+B0/40-(2/3-(B2-B0)/1200)*N),8
6280 FOR N = 0 TO 30 STEP 3
6282 PSET (220+N,20+B1/40-(2/3-(B2-B0)/1200)*N),8
6284 NEXT
6286 RETURN
```

PROG7

```
7000 'PROG7 ***
7001 'WE AND WS IN CM/DAY; H AND D IN METERS
7002 CLS:SCREEN 2:PRINT" ":PRINT" "
7004 PRINT "LPS MODEL WITH BUOYANCY FLUX ADDED"
7006 PRINT " ":PRINT ": PRINT "   STRIKE I TO CHANGE DEFAULTS[INITIALIZE]"
7008 PRINT "   STRIKE C TO CONTINUE[ACCEPTING DEFAULTS]"
7010 A$=INKEY$
7012 IF A$="C" THEN GOTO 7016
7014 IF A$="I" THEN GOTO 7022 ELSE GOTO 7010
7016 ' CONSTANTS AND PARAMTERS
7018 FAC = 333:YS0=30 :YB=30:WE3=-10 :WS2=-10:   'DEFAULTS CM/DAY
7020 GOTO 7030                              'CONTINUE
7022 INPUT "Y (LATITUDE OF SUBDUCTION [DEFAULT IS 30]";YS0
7024 INPUT "Y OF CHANGE OF SIGN OF BUOYANCY FLUX[DEFAULT WAS 30]";YB
7026 INPUT "W-EKMAN AT 30N     [ DEFAULT WAS -10 CM/DAY]";WE3:
7028 INPUT "W-INTERFACIAL AT 35N [ DEFAULT WAS -10 CM/DAY]";WS2
7030 X0 = -20:Y1=20:Y2=40:S1=.000003:S2=.000003:FAC=333
7032 BE=.00002:G=.01:D0=600:H0=0  'DEFAULT H0 AND D0 IN METERS
7034 TC= 5 :DC=10 :HC=50 :HDC=50 'CONTOUR INTERVALS IN METERS
7035 TC=7*TC:HC=7*HC:HDC=7*HDC:DC=7*DC
7036 LOCATE 22,1:CLS:SCREEN 1,0:COLOR 0,2:KEY OFF
7038 PI = 3.14159265# :FIRST=0
7040 LOCATE 22,1:'MAIN PROGRAM
7042 GOSUB 7070
7044 GOSUB 7096
7046 GOSUB 7110
7048 GOSUB 7160
7050 A$=INKEY$
7052 PRINT "STRIKE C TO CONTINUE"
7054 IF A$ = "C" THEN GOTO 7060
7056 GOSUB 7070
7058 GOTO 7050
7060 ' CHARACTERISTIC CALCS
7062 CLS
7064  GOSUB 7384
7066 'CONTINUATION OF MAIN PROGRAM
7068 GOTO  7196
7070 'BOXES
7072 LINE (0,0)-(100,100),1,B
7074 LINE (105,0)-(205,100),1,B
7076 LINE (210,0)-(310,100),1,B
7078 FOR I=0 TO 100
7080 X=X0+I*20/100
7082 L1=I:L2=105+I:L3=210+I
7084 M=100*(1-(YS0*COS(K0*(X-X0))-Y1)/20)
7086 PSET (L1,M),1:PSET(L2,M),2:PSET(L3,M),3
7088 NEXT
7090 M = 100*(1-(YB-Y1)/20)
7092 LINE (0,M)-(310,M),2,B
7094 RETURN
7096 ' WE AND WI  AND YS(X)
7098 YS=YS0*COS(K0*(X-X0))
7100 WE = (WE3*SIN(PI*(Y-Y1)/(Y2-Y1)))*BE*Y1^2/G : WE =  FAC*WE
7102 WI = -WS2 *(BE*Y1^2/G)*SIN(2*PI*(Y-Y1)/ (Y2-Y1)): WI=FAC*WI
7104 WEY = (WE3*COS(PI*(Y-Y1)/(Y2-Y1))*PI/(Y2-Y1))*BE*Y1^2/G:WEY=WEY*FAC
7106 IF Y>Y1 THEN GOTO 7108
7108 RETURN
7110 'WIND AND BUOYANCY CURVES
7112 LOCATE 14,12:PRINT "WE":LOCATE 14,27: PRINT "WS"
7114 PRINT" STRIKE C TO CONTINUE"
7116 M = 0
7118 L = 100+WE3*SIN(PI*M/100)
7120 PSET (L,M),3
7122 L1 = 210 + WS2* SIN(2*PI*M/100)
7124 PSET (L1,M),2
7126 M = M+1
7128 IF M<100 THEN GOTO 7118
7130 GOSUB 7376
7132 MS=100*(1-(YS-Y1)/20):MB=100*(1-(YB-Y1)/20)
7134 LINE(100,0)-(100,100),3
7136 L10=0:L30=210
7138 FOR I=0 TO 100
7140 L1=I:L3=210+I:X=X0+I*20/100
7142 MS=100*(1-(YS0*COS(K0*(X-X0))-Y1)/20 )
7144 LINE (L30,100)-(L3,MB),2,BF
7146 LINE (L30,MB)-(L3,MS),3,BF
```

```
7148 LINE (L10,100)-(L1,MS),2,BF
7150 LINE (L10,MS)-(L1,0 ),3,BF
7152 L30=L3 : L10=L1
7154 NEXT
7156 RETURN
7158 LOCATE 23,1: PRINT "WAIT FOR TRANSPORT CONTOURS TO BEGIN":'*******
7160 ' TRANSPORT CALCULATION
7162 X=X0:Y=Y1
7164 LOCATE 17,1: PRINT "WAIT FOR TRANSPORT CONTOURS TO BEGIN":'*******
7166 T2 = D0^2+2*WE*X
7168 T = SQR(T2)
7170 LOCATE 18,1: PRINT USING "SVERDRUP TRANSPORT ####.##";T-600
7172 L=(X-X0)*5+105:M=(Y2-Y)*5
7174 CT = 1 + INT(7*(T-D0)/TC) MOD 3
7176 PSET (L,M),CT
7178 B$=INKEY$
7180 IF B$="c" THEN GOTO 7060
7182 X = X+.3
7184 IF Y>=Y2 THEN RETURN
7186 IF X>=0 THEN GOTO 7190
7188 GOTO 7166
7190 X = X0:Y=Y+.3
7192 GOSUB 7096
7194 GOTO 7166
7196 ' CHAR FROM WEST, Y0>YS
7198 IF WE3=0 THEN GOTO 7328
7200 X0=-20:Y0=Y2-.1:X=X0:Y=Y0
7202 Y=Y0:X=X0
7204 B$=INKEY$
7206 IF B$="s" THEN GOSUB 7452
7208 Q0=300-(40-Y)*(290/20)
7210 LINE (Q0,110)-(300,190),0,BF
7212 GOSUB 7096
7214 D2=D0^2+2*WE*X0:D=SQR(D2):H=0
7216 GOSUB 7096
7218 UC =-WEY*X*Y: VC =   WE*Y
7220 X = X+S1*UC: Y=Y+S1*VC
7222 GOSUB 7230                    'DRAW OUT
7224 GOSUB 7096                    'NEED YS
7226 IF Y>=YS THEN GOTO 7216      'CONTINUE NORTH OF SUBDN.
7228 GOTO 7254      'AT OUTCROP
7230 ' COLOR CURVES
7232 LH = (X-X0)*5:LHD = LH+105: LD=LH+210
7234 M = (Y2-Y)*5
7236 CH = 1 + INT(7*H/HC) MOD 3
7238 'IF H=0 THEN CH=0
7240 CHD = 1 + INT(7*(D+H)/HDC) MOD 3
7242 CD=1+INT(7*  D    /DC) MOD 3
7244 IF LH>=0 AND H>0   THEN PSET(LH,M),CH
7246 IF LHD>=105 THEN PSET(LHD,M),CHD
7248 IF LD>=210  THEN PSET(LD,M),CD
7250 GOSUB 7356
7252 RETURN
7254 ' FIRST STEP AT OUTCROP
7256 S9=.000002 'FIRST STEP
7258 GOSUB 7096
7260 UC =-WEY*X*Y: VC =  WE*Y
7262 X=X+S9*UC:  Y = Y+S9*VC
7264 H = -S9*D*(WE-WI): IF H<0 THEN H=0
7266 D2=D0^2+2*WE*X-H^2: D=SQR(D2)
7268 GOSUB 7230                    'DRAW
7270 GOTO 7272
7272 'SUBDUCTED CHARACTERISTIC FROM WEST
7274 GOSUB 7096
7276 UC =-WEY*X*Y-(D-H)*H: VC=WE*Y
7278 X = X+S2*UC: Y=Y+S2*VC
7280 H = H + S2*(D*WI-(D-H)*WE)
7282 IF H<0 THEN H=0
7284 D2 = D0^2 +2*WE*X-H^2 :D=SQR(D2)
7286 IF D-H<=0 THEN GOTO 7296    'GET ANOTHER CHARACTERISTIC
7288 GOSUB 7230                    'DRAW
7290 GOSUB 7096                    'NEED NEW YS
7292 IF Z=1    THEN GOTO 7344
7294 IF X>X0 THEN GOTO 7274
7296 Y0=Y0-.3                      'STEP SOUTH
7298 Q0=300-(40-Y0)*(290/20)
7300 IF Y0>YS THEN GOTO 7202      'IF NORTH OF SUBDN, START AT TOP
7302 'RHINES-YOUNG                 'NEW CHAR SOUTH OF SUBDN--R/Y
```

153

PROGRAM 7

```
7304 Y=YØ : X=XØ
7306 GOSUB 7096
7308 D2MAX = DØ^2+2*WE*XØ
7310 DMAX=SQR(D2MAX)
7312 B=YØ*DMAX/YS   : C = B^2-D2MAX
7314 H =-B/2+SQR((B/2)^2-C): D = H+B
7316 LINE (QØ,110)-(300,190),Ø,BF        'ERASE DISPLAY
7318 UC =-WEY*X*Y-(D-H)*H
7320 IF UC<Ø THEN GOTO 7326
7322 GOTO 7272                           'CONTINUE WITH SUBDUCTED CHARACTERISTIC
7324 GOSUB 7202
7326 FIRST=Ø                             'START NEW CHARACTERISTICS FROM EAST WALL
7328 Z = 1:LINE (10,110)-(300,190),Ø,BF
7330 ' FIRST STEP
7332 X3= -.1  :YØ=20.2:D=DØ:Y=YØ :X=X3
7334 QØ1=300
7336 QØ=300-(40-YØ)*290/20
7338 GOSUB 7096
7340 H = SQR(2*WE*X)
7342 GOSUB 7230
7344 ' INTERIOR INTEGRATION
7346 IF X< -20 THEN GOTO 7352
7348 GOTO 7274
7350 IF X> -20 THEN GOTO 7344
7352 YØ=YØ+.5 : X3=-.1 : D=DØ:Y=YØ:X=X3:S=Ø:IF YØ>=YS THEN STOP
7354 GOTO 7338
7356 Q=300-(40-Y)*(290/20)
7358 PSET (QØ,109),Ø
7360 LINE(QØ,110)-(Q,198),Ø,BF
7362 LINE(QØ,110)-(Q,110+D/10),3,BF
7364 LINE(QØ,110+D/10)-(Q,110+D/10),CD
7366 IF H>Ø THEN LINE(QØ,110)-(Q,110+H/10),2,BF
7368 IF H>Ø THEN LINE(QØ,110+H/10)-(Q,110+H/10),CH
7370 QØ=Q:LOCATE 24,1: PRINT USING "WE###.# WS###.# CM/DY, H ### D ### M";WE/FAC
, WI/FAC,H,D:
7372 PSET (QØ,109),4
7374 RETURN
7376 B$=INKEY$
7378 IF B$="c" THEN GOTO 7382
7380 GOTO 7376
7382 RETURN
7384 'BOXES
7386 CH = 1 + INT(7*Ø/HC) MOD 3
7388 CD=1+INT(7* DØ   /DC) MOD 3
7390 LINE (Ø,Ø)-(100,100),1,B
7392 LINE (105,Ø)-(205,100),1,B
7394 LINE (210,Ø)-(310,100),1,B
7396 LO=Ø
7398 FOR I=Ø TO 100
7400 X=XØ+I*20/100
7402 M = 100*(1-(YS-Y1)/20)
7404 L1=I:L2=105+I:L3=210+I
7406 M=100*(1-(YSØ*COS(KØ*(X-XØ))-Y1)/20)
7408 PSET (L1,M),6:PSET(L2,M),6:PSET(L3,M),6
7410 LINE (LO,Ø)-(L1,M-1),CH
7412 LO=L1: NEXT
7414 M = 100*(1-(YB-Y1)/20)
7416 LINE (Ø,M)-(310,M),2,B
7418 LINE (Ø,Ø)-(100,Ø),CH
7420 LINE (100,Ø)-(100,100),CH
7422 LINE (100,100)-(Ø,100),CH
7424 LINE (210,Ø)-(310,Ø),CD
7426 LINE (310,Ø)-(310,100),CD
7428 LINE (210,100)-(310,100),CD
7430 DO=Ø
7432  FOR D=Ø TO 800 STEP 10
7434 CD=1+INT(7* D/DC) MOD 3
7436 LINE(305,110+DO/10)-(310,110+D/10),CD,BF
7438 DO=D
7440 NEXT
7442 HO=Ø
7444 FOR H=Ø TO 800 STEP 10
7446 CH=1+INT(7* H/HC) MOD 3
7448 LINE(1,110+HO/10)-(5,110+H/10),CH,BF
7450 HO=H: NEXT: RETURN
7452 A$ = INKEY$:'HERE AS HOLDING PATTERN
7454 IF A$="c" THEN RETURN ELSE GOTO 7452
7456 RETURN
```

154

PROGRAM 8

PROG8

```
8000 'PROG8 -NOTE LINE 8009 & 8238 FOR NEC OPERATION
8002 DIM XSH(150),YSH(150)              'SHADOW ZONE
8004   SCREEN 2
8006   KEY OFF
8008   KEY(11) ON :KEY(12) ON:KEY(13) ON:KEY(14) ON
8009 'NOTE: ADD 2 TO INKEYS FOR NEC APC
8010  SC=1000 ; YEAR=365*86400! :NY=1
8012 PI=3.14159
8014 WE=-.0000005:Y1=4000000!:Y0=1000000!:LW=PI/(Y1-Y0)
8016 BETA=2E-11:G=.01:D0=500:YS=3000000!
8018  DT=1000000!
8020 PRINT "SUBDUCTED THERMOCLINE MODEL"
8022 PRINT "TWO ACTIVE LAYERS-THE COLDER ONE IS COLOR 1 , THE WARMER IS"
8024 PRINT "COLOR 2.THE COLD LAYER OUTCROPS ONLY IN THE NORTH OF THE GYRE."
8026 PRINT "COLOR 3 IS THE EKMAN HAT."
8028 PRINT "THE NORTHERN BOUNDARY OF THE DOMAIN COINCIDES WITH WE=0."
8030 PRINT " "
8032 INPUT "ENTER LATITUDE OF SUBDUCTION OF LAYER 2[40>YS>10]";LS: LS=LS/10
8034 YS=LS*1000000!
8036 D02=D0^2 :FS=BETA*YS
8038 INPUT "ENTER 1 OR 2 FOR LAYER TO TRACE COLUMN IN [1 FOR WARM,2 FOR COLD]  "
;LA
8040   SCREEN 1:COLOR 0,2 :GOSUB 8212      'DRAW BOX
8042   GOSUB 8482             'WE PLOT
8044   GOSUB 8382             'SHADOW ZONE
8046 GOSUB 8412          'PLOT SHADOW ZONE
8048 L=40: IF LA=1 THEN M=100 ELSE M=25
8050 L9=L:M9=M:L8=L:M8=M:L7=L:M7=M
8051 GOSUB 8300
8052  GOSUB 8238
8054 'START
8056 DI=D:FI=F
8058 GOSUB 8212      'REDRAW BOX
8060 GOSUB 8412        'REDRAW SHADOW ZONE
8062 LOCATE 18,1
8064 'PRINT "RESTART CURSOR LAYER QUIT PRINT"
8066 'PRINT "    S      C      L      Q      P";
8068 REM INTEGRATE
8070 IF LA=1 THEN GOTO 8100
8072 IF Y<YS THEN GOSUB 8476            'REMOVE OLD VECTOR IN LAY 1
8074 GOSUB 8196               'GET VELOCITIES
8076 X=X+U2*DT :T=T+DT
8078 Y=Y+V2*DT
8080 A$=INKEY$
8082 IF A$="Q" THEN GOSUB 8448          'QUIT
8084 IF A$="P" THEN GOSUB 8422          'PRINT
8086 IF X<-3000000! THEN GOSUB 8448
8088 DE=DI*F/FI      'SHRUNK D FROM DI AT YI[FI]
8090 DS=DI*F/FS
8092 GOSUB 8228
8094 GOSUB 8356
8096 GOSUB 8516
8098 GOTO 8074
8100 T=0:HI=H:FI=F:DI=D
8102 GOSUB 8204
8104 X=X+U1*DT:Y=Y+V1*DT:T=T+DT:B=H+D
8106 A$=INKEY$
8108 IF A$="Q" THEN GOSUB 8448          'QUIT
8110 IF A$="P" THEN GOSUB 8422          'PRINT
8112 HE=HI*F/FI
8114 IF X<-3000000! THEN GOSUB 8448
8116 HE=HI*F/FI
8118 GOSUB 8228
8120 GOSUB 8356
8122 GOSUB 8516
8124 GOTO 8102
8126 'Y>=YS CASE
8128 GOSUB 8346      'GET W,WY,F
8130 D2=D02+2*W*X      'D*D
8132 D =SQR(D2)
8134 U2=-G*WY*X/(F*D )
8136 V2= G*W/(F*D )
8138 H =0
8140 U1=U2
8142 V1=V2
8144 DS=D             'UPDATE SUBDUCTION VALUE EACH TIME
```

155

```
8146 IF LA= 1 THEN BEEP: GOTO 8004
8148 RETURN
8150 'Y<YS    SUBDUCTED CASE
8152 GOSUB 8346
8154 F2=1+(1-F/FS)^2
8156 D2=(D02+2*W*X)/F2              'D*D
8158 IF D2 >= D02 THEN GOTO 8180    'NOT IN SHADOW D^2<=D0^2 ZONE
8160 D =D0:U2=0:V2=0
8162 H2= 2*W*X              'H*H
8164 IF H2<=0 THEN H =0 ELSE GOTO 8172    'TROUBLE
8166 REM TROUBLE
8168 U1=0:V1=0
8170 RETURN
8172 H =SQR(H2)
8174 U1=-(G*WY*X)/(F*H )
8176 V1= (G*W)/(F*H )
8178 RETURN
8180 D =SQR(D2)
8182 H =(1-F/FS)*D
8184 B =H +D
8186 U2=-(G/F)*(WY*X+BETA*(1-F/FS)*D*D/FS)/(F2*D)
8188 V2= G*W/(F*F2*D )
8190 U1=(2-F/FS)*U2+(G*BETA*D )/(F*FS)
8192 V1=(2-F/FS)*V2
8194 RETURN
8196 'LAYER 2
8198 IF Y>=YS THEN GOSUB 8126
8200 IF Y<YS THEN GOSUB 8150
8202 RETURN
8204 'LAYER 1
8206 IF Y>YS THEN BEEP:GOTO 8038
8208 GOSUB 8150
8210 RETURN
8212 LINE (150,10)-(0,160),3,B
8214 LINE (0,5)-(300,5),3
8216 FOR I=0 TO 300 STEP 10
8218 LINE (I,0)-(I,6),3
8220 NEXT
8222 M =160-50*((YS-Y0)/1000000!)
8224 LINE (0,M )-(150,160),3,B
8226 RETURN
8228 REM CONVERT TO PLOT SCALES
8230 L9=150+50*(X/1000000!):M9=160-50*((Y-Y0)/1000000!)
8232 IF LA=1 THEN CC=2 ELSE CC=1
8234 PSET (L9,M9),CC
8236 RETURN
8238 ' CURSOR KEYS; NOTE: ADD 2 TO KEYS IF USING NEC APC
8240 ON KEY(11) GOSUB 8256     'UP
8242 ON KEY(12) GOSUB 8264      'LEFT
8244 ON KEY(13) GOSUB 8272     'DOWN
8246 ON KEY(14) GOSUB 8280     'RIGHT
8248 A$=INKEY$
8250 IF A$="G" THEN GOTO 8054   'START COMPS
8252 IF A$="" THEN GOTO 8238
8254 RETURN
8256 'CURSOR UP
8258 M=M-1
8260 GOSUB 8300
8262 RETURN
8264 'CURSOR LEFT
8266 L=L-1
8268 GOSUB 8300
8270 RETURN
8272 'CURSOR RIGHT
8274 L=L+1
8276 GOSUB 8300
8278 RETURN
8280 'CURSOR DOWN
8282 M=M+1
8284 GOSUB 8300
8286 RETURN
8288 'TRANSFORM L9,M9 TO X,Y
8290 X=1000000!*(L-150)/50:Y=Y0+1000000!*(160-M)/50
8292 RETURN
8294 'X,Y TO L,M
8296 L=150+50*X/1000000!:M=160-50*(Y-Y0)/1000000!
8298 RETURN
8300 ' PLOT
8302 LINE (L9,M9)-(L8,M8),0
```

PROGRAM 8

```
8304 IF YO<YS THEN LINE (L9,M9)-(L7,M7),0
8306 CIRCLE (L9,M9),2,0
8308 PAINT (L,M),0
8310 GOSUB 8288         'GET X,Y
8312 GOSUB 8336         'GET U,V
8314 L1=L+SC*U2:M1=M-SC*V2
8316 L2=L+SC*U1:M2=M-SC*V1
8318 'PSET (L,M),0
8320 IF LA=1 THEN CC=2 ELSE CC=1
8322 CIRCLE (L,M),2,CC        'CENTER
8324 PAINT (L,M),CC
8326 LINE (L,M)-(L1,M1),1
8328 IF Y>YS THEN GOTO 8330 ELSE LINE (L,M)-(L2,M2),2
8330 L9=L:M9=M:L8=L1:M8=M1:L7=L2:M7=M2:YO=Y      'OLD VALUES
8332 RETURN
8334 'REM
8336 'GET VELOCITY AND DEPTHS
8338 'GET DEPTHS AND VELOCITIES
8340 IF Y>=YS THEN GOSUB 8126
8342 IF Y<YS THEN GOSUB 8150
8344 RETURN
8346 FAC=BETA*WE/G
8348 W=(BETA*Y*Y/G)*WE*SIN(LW*(Y-Y0))
8350 WY=FAC*Y*Y*LW*COS(LW*(Y-Y0))+2*W/Y
8352 F=BETA*Y
8354 RETURN
8356 'PLOT D,H BARS
8358 L5=175+D/10:L6=175+(D-H)/10:L7=175+DE/10
8360 IF LA=1 GOTO 8374
8362 LINE (175,M9)-(L5,M9),1
8364 IF Y>YS THEN LINE (L7,M9)-(L5,M9),8 : GOTO 8372
8366 LL=175+DE/10: LL1=175+DS/10
8368 LINE (LL,M9)-(LL1,M9),8
8370 LINE (L6,M9)-(L5,M9),2
8372 RETURN
8374 LINE (175+H/10,M9)-(L6+H/10,M9),1      'LOWER LAYER
8376 LINE (L6+H/10,M9)-(175+H/10+(D-H+HE)/10,M9),2 'NON-EKMAN HAT
8378 LINE (175+H/10+(D-H+HE)/10,M9)-(L5+H/10,M9),8 'EKMAN HAT
8380 RETURN
8382 'SHADOW ZONE
8384 FOR I=1 TO 150
8386 IF X<-100000! THEN GOTO 8394
8388 Y =YS-10*(YS-Y0)*(I-1)/149
8390 YO=Y
8392 GOTO 8396
8394 Y =YO-(YS-Y0)*(I-1)/149
8396 GOSUB 8346
8398 X =D02*((1-F/FS)^2)/(2*W)
8400 IF X< -3000000! THEN GOTO 8408
8402 GOSUB 8294        'GET LM
8404 XSH(I)=L:YSH(I)=M
8406 NEXT
8408 NSH=I
8410 RETURN
8412 'GET SHADOW ZONE PLOT
8414 FOR I=1 TO NSH
8416 PSET (XSH(I),YSH(I)),3
8418 NEXT
8420 RETURN
8422 'PRINT OUT RESULTS
8424 IF LA=2 THEN PRINT USING "####.###";D,H,U2,V2,T/YEAR:RETURN
8426  PRINT USING "####.###";D,H,U1,V1,T/YEAR
8428 RETURN
8430 'LAYER INTERFACES-GIVEN Y
8432 GOSUB 8346        'GET W,WY,F
8434 FOR I=1 TO 100
8436 X=-3000000!*(I-1)/999
8438 IF Y>YS THEN GOTO 8446
8440 D=SQR(D02+2*W*X)
8442 H=0
8444 STOP
8446 D2=(D02+2*W*X)/(1+(1-F/FS)^2)
8448 'REDO WITH ANOTHER STARTING POINT
8450 A$=INKEY$
8452 IF A$="" THEN GOTO 8450
8454 IF A$="L" THEN GOTO 8462
8456 IF A$="S" THEN GOTO 8004
8458 IF A$="C" THEN GOTO 8468
8460 RETURN
```

157

```
8462 ' RESET LAYER
8464 IF LA=1 THEN LA=2 ELSE LA=1
8466 GOTO 8468
8468 'REDO WITH CURSOR
8470 IF LA=1 THEN M=130 ELSE M=25:L=25
8472 GOSUB 8504          'REMOVE BARS OR WHATEVER
8474 GOSUB 8238
8476 'REMOVE OLD VECTOR IN LAYER 1
8478 LINE (L9,M9)-(L7,M7),0
8480 RETURN
8482 'PLOT WE
8484 LINE (310,10)-(310,160),1
8486 FOR I=1 TO 150
8488 Y=Y1-(Y1-Y0)*(I-1)/149
8490 W9=WE*SIN(LW*(Y-Y0))
8492 W8=W9*(((Y-Y0)/(Y1-Y0))^2)
8494 L9=310+W9*5E+07:M9=160-50*((Y-Y0)/1000000!)
8496 L8=310+W8*5E+07
8498 PSET (L9,M9),3 :PSET (L8,M9),2
8500 NEXT
8502 RETURN
8510 LINE (175,10)-(250,160),0,B
8512 PAINT (200,75),0
8514 RETURN
8516 'TIME LINE
8518 LT=T/YEAR
8520 LINE (LT,4)-(LT,2),6,B
8522 IF LT > NY*10 THEN NY=NY+1: BEEP
8524 RETURN
```

PROG9

```
9000 'PROG9
9002 CLS: KEY OFF: SCREEN 1: COLOR 0,2
9004 LINE (120,30)-(140, 50),1,BF
9006 LINE(120, 60)-(140, 80),2,BF
9008 LINE(120, 90)-(140,110),3,BF
9010 LINE(120,120)-(140,140),8,BF
9012 LOCATE 6,5 :PRINT "COLOR 1"
9014 LOCATE 10,5:PRINT "COLOR 2"
9016 LOCATE 14,5: PRINT "COLOR 3"
9018 LOCATE 18,5: PRINT "COLOR 8"
9020 LOCATE 2,10: PRINT" COLOR CODE "
9022 END
```

Glossary

advection: the process in which properties of a fluid are transported by the fluid velocity rather than by diffusion, sinking, etc.

beta-spiral: the rotation with depth of the direction of slopes of density surfaces and their accompanying geostrophic velocity in mid-ocean current regimes.

buoyancy boot: the amount of water added to bottom of a water column due to flux of water across a density interface.

characteristic curve: a mathematical curve along which the integration of a differential equation can be performed.

Coriolis force: a fictitious force that is introduced into the dynamical equations of motion when referred to a uniformly rotating reference frame such as the earth, as a convenient representation of the accelerations encountered by parcels of fluid or particles moving with respect to the overall rotational equilibrium of the earth.

Ekman hat: the amount of water added to the top of a water column in the upper layer of the ocean by downward pumping of water by the wind stress.

Ekman layer: a thin (100 meter) layer at the top of the ocean driven primarily by the wind stress. Its transport is generally to the right of the wind stress in the northern hemisphere.

Ekman pumping: the forcing downward of water into the layers beneath the Ekman layer in regions where the horizontal transport of the Ekman layer tends to accumulate mass, as is the case over much of the subtropical ocean.

Ekman transport: the vertically integrated horizontal transport of the Ekman layer, proportional to the wind stress divided by sine of latitude, and directed to the right of the wind stress in northern hemisphere.

geostrophic velocity: the horizontal velocity of fluid on the earth that provides the necessary Coriolis force to balance a local horizontal pressure gradient.

gyre: the semi-elliptical pattern of ocean circulation observed over major ocean basins. In the northern subtropics the sense of rotation is clockwise, and several decades are required to complete a circuit.

horizontal pressure gradient: the change of pressure in the horizontal direction. In hydrostatic situations it results from variations in atmospheric pressure at sea level, in the level of the sea surface and in the depth of density surfaces inside the ocean.

hydrostatic equilibrium: perfect balance at all points within a fluid between the local acceleration of gravity downward and the upward vertical pressure gradient.

indirect buoyancy convection: a common oceanic situation in which rising and warming water flows toward the equator rather than toward the poles.

isobar: a curve of a level surface along which the pressure is constant.

Luyten algorithm: a set of computational rules for determining layer depths in a model ocean driven by both wind and buoyancy forcing.

overdetermined: used in referring to a set of mathematical statements that out-number the unknown variables, and hence may be contradictory.

potential thickness: a fictitious thickness computed by dividing the real thickness of a density layer by the sine of latitude. In simple oceanic current systems, the flow is expected to be along contours of equal potential thickness.

rain hat: the amount of water added to the top of a column of water by the excess of precipitation over evaporation.

recirculation: the number of complete circuits that a particle of water makes around a gyre from its time of entry in a layer to its time of exit.

reference layer: one of a set of superposed density layers, usually the bottom layer, in which provisional estimates of horizontal pressure gradient (and correspond-ing geostrophic velocity) are made in order to compute the velocities in the other layers. These reference values are sometimes determinable by making a diag-nostic calculation.

Rossby wave: long Rossby waves are systems of geostrophic currents that propa-gate westward in an otherwise resting ocean, due to a periodic shrinking and stretching of water columns oscillating northward and southward.

Rossby repeller: a point or line in the ocean that characteristic curves avoid.

shadow zone: portion of a density layer in an oceanic gyre that is not directly ac-cessible to water forced downward at subduction fronts.

subduction: the process by which water in a surface layer is led down under an out-cropping density interface to feed the subsurface circulation elsewhere in the gyre.

Sverdrup transport and velocity: The vertically integrated transport in the ocean at any point in the ocean can be expressed in terms of the local torque of the wind stress, a fact discovered by Sverdrup. When this is divided by an appropriate depth scale, we can speak of a Sverdrup velocity.

underdetermined: used in referring to a set of mathematical statements that are outnumbered by the unknown quantities and therefore insufficient to determine them uniquely.

Select Bibliography

Armi, L., and H. Stommel. 1983. Four Views of a Portion of the North Atlantic Subtropical Gyre. *Jour. Phys. Oceanog.* 13: 828-857.

Behringer, D., and H. Stommel. 1980. The Beta-Spiral in the North Atlantic Subtropical Gyre. *Deep-Sea Res.* 27a: 225-238.

Jenkins, W. 1986. ^{3}H and ^{3}He in the Beta Triangle; Observations of Gyre Ventilation and Oxygen Utilization rate. *Jour. Phys. Oceanog.* (in press).

Luyten, J., J. Pedlosky, and H. Stommel. 1983. The Ventilated Thermocline. *Jour. Phys. Oceanog.* 13: 292-309.

Luyten, J., and H. Stommel. 1986. Gyres Driven by Combined Wind and Buoyancy Flux. *Jour. Phys. Oceanog.* 16: 1551-1560.

Pivar, M., E. Fredkin, and H. Stommel. 1963. Computer Compiled Oceanographic Atlas: An Experiment in Man-Machine Interaction. *Proc. Nat. Acad. Sci. (U.S.)* 50: 36-38.

Rhines, P., and W. Young. 1981. A Theory of the Wind-driven Circulation. *J. Mar. Res.* 40 (suppl. 1): 559-596.

Schott, F., and H. Stommel. 1978. Beta-Spirals and Absolute Velocities in Different Oceans. *Deep-Sea Res.* 25: 961-1010.

Veronis, G. 1978. Model of world ocean circulation, III: Thermally and wind driven. *J. Mar. Res.* 36: 1-44.

Index

quarantine of *Beagle*, 22

rain hat, 54
reference level, 37, 56; layer, 45
relative horizontal pressure gradient, 37
relative vorticity, 68, 73, 83
Rhines, P., xiii
Rossby: attractor, 99; repeller, 99; wave, 92, 98
rotation: differential, 19; of earth, 13ff.; solid, 16; of velocity with depth, 64
rules, first three, 44-45; fourth, 90; fifth, 91; exceptions to, 68ff.; table with codicils, 69

salinity of North Atlantic, 117
satellite, promise of, 11, 35; heat flux from, 71
Schott, F., 62, 65
sea-surface level, unknown, 11
sections, vertical, 5; North Atlantic, 116
shadow zone, 79
sigma-t, definition of, 120
solenoids, parallel, 51
South Atlantic, 123
spheres, concentric, 5
spin, 68, 72
Spitzer, L., 3
Stanley, R., 22
station, test, 31
stationary flow patterns, 21
stream tube, 40-41
string and stone, 24
subduction, 79, 99, 105, 107, 123
subtropical gyre, 64
Sverdrup, H., 90

Sverdrup experiment, 62, 90-93
Swallow, J., 114

temperature of North Atlantic, 117
Tenerife, 2
Tenner, E., xiv
thermocline, 17; ventilated, 93
thickness curtain, 40
Toulon, 22
Trades, 31, 70
trajectory of cloud, 122
tritium and helium-3, xiii, 33, 66
tsunamis, 9
tuning spiral to latitude, 65
twisting of direction, 52

undetermined systems, 51
upper-air observations, 71

ventilation, 78, 93, 123
Veronis, G., 92-93
Veronis equation, 91
vertical, acceleration, 6; scale exaggeration, 5
Von Neumann, J., 93

wall of warm water, 27
warm core, 85
watches at sea, 33
water columns, 36
waves, from bumps, 9
westerlies, 31, 70
wind hat, 55
wind-driven layer, 63, 70
Woods Hole, 22, 63; Oeanographic Institution, 3, 85
Worthington, V., 114

Yeats, W., 112
Young, W., xiii

Library of Congress
Cataloging-in-Publication Data

Stommel, Henry M., 1920-
A view of the sea.

Bibliography: p.
Includes index.
1. Ocean circulation. I. Title.

GC228.5.S75 1987   551.47   86-30502
ISBN 0-691-08458-0 (alk. paper)